BAOFENG RADIO SURVIVAL HANDBOOK

Essential Communication Skills to Stay Safe During Emergencies, Unforeseen Disasters, and Extreme Outdoor Activities

ALEX RANGER

Table of Content

Introduction

Why Baofeng Radio

When it comes to communication tools, Baofeng radios have gained a lot of popularity due to the combination of a wide range of features and practical benefits that together with the price tag resonate well with its target audience.

Baofeng radios are produced by the Chinese manufacturer Baofeng and it's most popular model is the UV-5R. It is a dual-band FM radio able to transmit and receive on both VHF (Very High Frequency) and UHF (Ultra High Frequency) bands. This makes it very versatile and suitable for different communication needs and scenarios, from short-range outdoor excursions to long-range emergency situations.

Its good performance and versatility together with the accessible price tag make Baofeng very cost-effective, making it the first choice for a broad range of individuals. In fact, Baofeng usually recommended to and often adopted by first-time buyers as these models are renowned for their user-friendly interfaces, making them accessible to both newcomers and seasoned operators. Simplified programming options and intuitive controls ensure a quick grasp of radio communication's bases.

In addition to its good performance and affordable price, portability is another strength of this radio. Baofeng radio is also indicated for those who are constantly on the move. Its designs are created with convenience in mind, aiming for a compact but lightweight body which is easy to carry around.

Baofeng performs well during emergencies offering reliable communication and excelling in providing a robust connection even in adverse conditions, when conventional communication systems fail.

Even in terms of customization and programming, Baofeng radios offer a lot of flexibility, empowering users to tailor their communication experience. By selecting specific frequencies, setting up channels, and configuring settings, individuals can optimize their radios for distinct purposes, whether it's for outdoor adventures, emergency preparedness, or amateur radio operations.

From a technical perspective instead, these are the main features Baofeng radios include:

Feature	Description
Frequency Range	VHF: 136-174 MHz / UHF: 400-520 MHz
Output Power	1 watt / 4 watts (selectable)
Dual-Band Capability	Yes
FM Radio	Yes
Channel Capacity	128 programmable channels
Display and Controls	LCD display with keypad

CTCSS and DCS	Supported
VOX (Voice-Activated Transmission)	Yes
Dual Watch and Dual Reception	Yes
Battery	Rechargeable lithium-ion battery pack
Antenna	Removable antenna, upgradeable to higher-gain
Squelch and Scan Functions	Yes

Here are some additional details about the different characteristic of the Baofeng UV-5R:

Overall feature	Details
Design and build	The UV-5R is compact and lightweight and it usually features a rugged design. It's built to withstand typical outdoor conditions, making it suitable for various activities such as hiking, camping, and emergency communications.
Display and User Interface	The radio comes with an LCD display that shows essential information - selected channel, frequency, battery status, signal strength, and other operational parameters. The user interface normally includes a numeric keypad and function keys for easy navigation and programming.
Programming	The UV-5R can be programmed manually through the keypad, allowing users to set frequencies, tones, and other parameters for each channel. Programming software and a programming cable are also available, enabling users to program the radio using a computer, which makes things easier when managing large channel lists.
Dual Standby	The radio supports dual standby, meaning it can monitor two channels at the same time. When a transmission is received on one channel, the UV-5R automatically switches to the other channel, allowing users to respond to incoming calls promptly.
Power Options	The Baofeng UV-5R typically comes with a rechargeable lithium-ion battery pack, providing several hours of use on a single charge. There are various battery options available, including extended-life batteries for longer operating times. It can also be powered using AA batteries using a battery case adapter.
Antenna	The radio is equipped with a removable antenna, which is needed for its dual-band capability. The standard antenna can be replaced with higher-gain antennas to increase the communication range.
Accessories	There are many accessories available for the UV-5R - headsets, speaker microphones, carrying cases, and external antennas.
Licencing and regulations	As with any two-way radio, it's essential to comply with the radio communication regulations and obtain the necessary licenses before using the Baofeng UV-5R on certain frequencies or in certain regions. Users are strongly encouraged to research the regulations in vigor in their area.
Amateur use	For licensed amateur radio operators, the Baofeng UV-5R provides access to a wide range of frequencies and bands, allowing them to communicate with other hams locally and globally.
Firmware and	The Baofeng UV-5R has undergone several revisions since its initial

versions	release, resulting in different firmware versions and sub-models. These revisions included bug fixes, specific improvements, or changes in features. When purchasing the radio, it's advisable to check for the latest version and firmware available.

It's important to note that the UV-5R, like many Baofeng radios, is primarily intended for amateur radio use and personal communication needs. Always be aware of the laws and regulations governing the use of two-way radios in your region and follow proper radio etiquette while communicating to ensure smooth and interference-free communication with others.

The final perk of this model is its extended availability. The UV-5R model, is easily available through online retailers and various marketplaces. Its visibility and availability have contributed to its popularity.

Cons

After having highlighted the main reasons Baofeng radios and the UV-5R model in particular are highly popular in the amateur market, it seems necessary to provide a quick overview of the negative sides of these products.

Cons	Descriptions
Build quality	Baofeng radios may have lower build quality compared to more expensive brands, making them less durable.
Interference	These radios may be more susceptible to intermodulation and emissions, leading to potential interference.
User interface	The menu navigation can be challenging, especially for beginners.
Battery life	Battery life may not be as long-lasting compared to higher-priced radios.
Lack of support	Baofeng radios may not come with comprehensive customer support or warranties compared to established brands.
Counterfeit Products	Due to their popularity, counterfeit Baofeng radios have been known to circulate, leading to potential issues.
Non-Compliance	Some versions of Baofeng radios may not meet specific radio regulations in certain countries, leading to restrictions.

Despite the above, Baofeng radios and the UV-5R model remain a favorite, reliable options for amateurs around the world.

Personal Radio Services

Summary of Personal Radio Services and Baofeng Compatibility in the USA:

In the United States, Personal Radio Services refer to various radio communication services that are intended for personal or business use, outside the realm of traditional commercial broadcasting. These services provide convenient and flexible communication options for individuals, families, businesses, and organizations. There are several types of Personal Radio Services in the USA, each with its own set of rules and regulations.

Citizens Band (CB) Radio: CB radio is a popular Personal Radio Service that operates on 40 specific channels within the 27 MHz band. It is commonly used by truckers, off-road enthusiasts, and travelers to communicate over short distances. Baofeng radios are not compatible with CB frequencies as they operate on different frequency bands.

Family Radio Service (FRS): FRS is a license-free service that offers short-range communication over 22 channels in the UHF band. It is ideal for recreational activities, family outings, and small businesses. Baofeng radios, such as the UV-5R series, can access FRS frequencies, but it's essential to comply with power output limitations to avoid violating FCC regulations.

General Mobile Radio Service (GMRS): GMRS requires a license from the FCC and provides higher power and longer-range communication compared to FRS. It offers 30 channels, including repeater channels, and is commonly used for outdoor activities, emergency communication, and businesses. Baofeng radios, like the BF-F8HP, are compatible with GMRS frequencies and can be used with a valid GMRS license.

Multi-Use Radio Service (MURS): MURS is a license-free service with five VHF channels, designed for short-range communication in personal and business applications. Baofeng radios, including the UV-5R, can access MURS frequencies without a license, making them a suitable choice for MURS communication.

Family Radio Service (FRS) and General Mobile Radio Service (GMRS) Hybrid Radios: Some Baofeng radios, such as the UV-82C, are marketed as FRS/GMRS hybrid radios. While these radios can access both FRS and GMRS frequencies, it's essential to note that a valid GMRS license is required to use GMRS channels legally.

It is crucial to comply with FCC regulations and licensing requirements when using Baofeng radios or any other personal radio devices. Operating on unauthorized frequencies or using higher power levels without a valid license can result in severe penalties and interference with other radio users.

Before using Baofeng radios for personal or business communication, it is advisable to familiarize yourself with the specific regulations governing each type of Personal Radio Service. Obtain the necessary licenses, adhere to power limitations, and always practice responsible and ethical radio use to ensure a smooth and efficient communication experience.

Chapter 1: Introduction to reliable communication

Exploring Fundamental Radio Communication Concepts and Terminology

Radio communication is a way to transmit information across long distances in our modern society. If you are interested in exploring the world of radios, it is crucial to understand the fundamental concepts and technical terms tied to radio waves, including electromagnetics. This section will delve into the fundamental principles that form the foundation of waves and radio communication.

Radio Waves and Electromagnetic Spectrum

The electromagnetic spectrum composes the entirety of electromagnetic radiation, also called electromagnetic waves, each characterized by unique frequencies and wavelengths. These waves are fundamental to various technologies and have diverse applications across multiple fields, including communication, imaging, and scientific research.

Properties of Electromagnetic Waves

All electromagnetic waves share common properties:

- **Speed:** All electromagnetic waves travel at the speed of light in a vacuum, denoted as "c," approximately 299,792,458 meters per second.

- **Waveform:** They exhibit periodic oscillations of electric and magnetic fields perpendicular to each other and the direction of wave propagation.

- **No Medium Required:** Electromagnetic waves can travel through a vacuum as well as through materials with different properties. e.g. water, air, metals (solids).

- **Quantization:** This is a topic that we will not delve in this book, but without going into too much detail (and without opening a can of worms), understand that according to quantum mechanics, electromagnetic waves can act both as waves (wave-particle duality) and particles (photons).

Region	Frequency Range (Hz)	Wavelength Range (meters)	Applications
Gamma Rays	$> 10^{19}$	$< 10^{-12}$	Nuclear research, medical imaging (PET scans)

X-rays	10^{16} - 10^{19}	$10^{(-12)}$ - 10^{-9}	Medical imaging, security scanning, astronomy
Ultraviolet (UV)	10^{15} - 10^{16}	10^{-9} - 10^{-7}	UV sterilization, fluorescence, astronomy
Visible Light	4.3×10^{14} - 7.5×10^{14}	7.0×10^{-7} - 4.0×10^{-7}	Human vision, optical communications, spectroscopy
Infrared (IR)	3×10^{11} - 4.3×10^{14}	$1 \times 10^{(-3)}$ - 7.0×10^{-7}	Thermal imaging, remote controls, communication
Microwaves	10^{9} - 3×10^{11}	1×10^{-1} - 1×10^{-3}	Radar, satellite communication, microwave ovens
Radio Frequencies	< 10^{9}	> 1×10^{-1}	Radio broadcasting, wireless communication, RFID

Radio waves are a only a portion of the electromagnetic spectrum and occupy the lowest frequency range. This range goes is approximately 3kHz – 300GHz. Similarly, the wavelength for radio waves is approximately 100 kilometers (km) to 1 millimeter (mm). We can interchangeably use the terms "radio waves", "radio frequencies" or "radio signals" to refer to the same thing.

Engineers use the relationship between frequency and wavelength to optimize radio systems for specific use cases. Here are some common frequency bands and their corresponding wavelength ranges for radio waves:

Frequency Range (Hz)	Wavelength Range (meters)	Typical Applications
3 kHz - 30 kHz	100 km - 10 km	Long-range radio communication
30 kHz - 300 kHz	10 km - 1 km	Maritime communication (Medium wave)
30 kHz - 300 kHz	10 km - 1 km	Maritime communication (Medium wave)
300 kHz - 3 MHz	1 km - 100 m	AM radio broadcasting
30 MHz - 300 MHz	10 m - 1 m	VHF TV and radio broadcasting
3 GHz - 30 GHz	10 cm - 1 cm	Microwave communication and radar

As radio waves have the ability to travel long distances and penetrate obstacles with minimal attenuation, it makes them well-suited for various communication applications, particularly in wireless technology and broadcasting.

Frequency and Wavelength

As aforementioned, the frequency and wavelength determine the characteristics of an electromagnetic wave:

Frequency: Frequency (**f**) refers to the number of cycles (or oscillations) that occur in one second. It is measured in Hertz (Hz), where 1 Hz is equal to 1 cycle/second.

Wavelength: Wavelength corresponds to the distance between two points in an electromagnetic wave that are in phase (such as two peaks or two troughs). It is denoted by the Greek letter lambda (**λ**) and the usual unit of length it is measured in, is in meters (m).

Relationship between Frequency and Wavelength:

Frequency (**f**) and wavelength (**λ**) are inversely proportional to each other in a vacuum or air. This relationship is described by the equation:

$$c = f * \lambda$$

Where:

c is the speed of light in a vacuum (approximately 299,792,458 meters per second)

Frequency and wavelength have an inverse relationship (as one increases the other decreases). This ensures that the product of frequency and wavelength remains constant at the speed of light.

Radio Working Principle

The generation of radio frequencies involves creating oscillating electrical currents that produce alternating electric and magnetic fields. These fields propagate as radio waves, which can travel through the air or other media, such as cables or waveguides. The main journey of a radio communication is:

1. **Transmitter:** The transmitter modulates the information (voice, data, etc.) onto a carrier signal by varying its amplitude, frequency, or phase. The resulting modulated signal contains the desired information.

2. **Antenna:** Antennas receive the electrical signal (modulated signal) and convert this into radio waves. They are designed to radiate these waves efficiently into space or a specific direction.

3. **Propagation:** The radio waves travel through the air or a medium until they reach the intended receiver.

4. **Receiver (antenna):** The receiver's antenna captures the radio waves and converts them back into electrical signals.

5. **Demodulation:** The received signal is then demodulated to extract the original information, which can be heard, seen, or processed by the intended user.

Applications of Radio Frequencies

Radio waves find applications in numerous technologies, some of which are:

1. **Broadcasting:** AM (Amplitude Modulation) and FM (Frequency Modulation) radio are common examples of broadcasting. These are the main two that come to mind when we think "radio".

2. **Telecommunications:** Mobile phones, Wi-Fi, Bluetooth, and satellite communication systems utilize radio frequencies.

3. **Radar:** Radar systems use radio waves to detect and track objects, used in aviation, weather forecasting, and military applications.

4. **Wireless Data Transmission:** Wi-Fi and cellular networks enable wireless data communication.

5. **Radio Astronomy:** Radio frequencies are used to study celestial objects in space.

Modulation and Demodulation

In radio communication, modulation is the process of impressing information onto a carrier signal. The carrier signal acts as a conveyance for the information to be transmitted. Here I will introduce two terms you may be familiar with, AM and FM. "Amplitude Modulation" (AM) and "Frequency Modulation" (FM) are two different modulation techniques used to modulate the carrier signal. Another less commonly known modulation technique is Single Sideband (SSB) modulation.

Demodulation describes the process of extracting the original information from the modulated carrier signal at the receiving end. The demodulator separates the information from the carrier signal, allowing the receiver to recreate the original message.

Transmitters and Receivers

Transmitters and receivers are fundamental components of any radio communication system. The transmitter is responsible for converting the audio or data signals into modulated radio waves suitable for transmission. On the other end, the receiver detects and demodulates the received radio waves to extract the original information.

Antennas

As you can imagine, Antennas are devices designed to transmit or receive radio waves efficiently. They serve as conduits for propagating electromagnetic waves (radio waves) between the transmitter and receiver. Depending on the specific characteristics of the antenna (shape, size), they can be used for specific needs (frequencies/wavelength) and as such, different applications. Proper antenna selection and positioning are crucial to achieve optimal communication performance.

Channel and Bandwidth

A channel in radio communication refers to a specific frequency range allocated for transmission. Wider bandwidth allows for higher data rates, but also requires a wider frequency spectrum.

Think of the channel as a highway, and the bandwidth is its number of lanes. The more lanes (or broader the bandwidth), the more information can travel through at the same time. So, in layman terms, the bandwidth of the channel is like its capacity, determining how much information it can carry within its frequency range.

Line-of-Sight and Beyond Line-of-Sight Communication

As all electromagnetic waves, radio waves travel in straight line and propagate until they encounter an obstacle. Line-of-sight (LOS) communication happens when there is a clear path between the transmitter and receiver. Beyond line-of-sight communication involves using techniques like reflection, diffraction, and refraction to overcome obstacles and reach distant locations.

Signal Propagation and Attenuation

As radio signals, they are subject to various factors that can affect their strength and quality. Signal attenuation refers to the weakening of a signal as it travels through the objects. Obstructions, distance, atmospheric conditions, and interference contribute to signal degradation.

Unveiling the Advantages of Ham Radio for Outdoor Enthusiasts and Preppers

"Ham radio" is the term used commonly to describe an amateur radio which is used to exchange non-commercial messages. This is the main choice for outdoor enthusiasts and preppers. Ham radio operators posses several advantages over other radio communication alternatives, making it a solid option for many scenarios. In this section, we will look at the benefits that amateur radios (Ham radio) offer, and why it is a preferred option for those venturing into the outdoors or preparing for emergencies.

Independence from Infrastructure

One of the main advantages of Ham radio is its ability to function independently of traditional communication infrastructure. Unlike cell phones or internet-based communication, Ham radio operators do not rely on cellular towers, internet connectivity, or any centralized system. This self-reliant nature enables communication even in remote and isolated areas where other forms of communication may not be available during emergencies or outdoor adventures.

Long Communication Range

Ham radio operates across a wide range of frequencies, including very high frequency (VHF) and ultra-high frequency (UHF) bands. These frequencies allow for reliable long-distance communication, especially in favorable conditions with a clear line of sight. Ham radio enthusiasts often employ repeater stations, which retransmit signals over longer distances, further extending their communication range.

Interoperability and Emergency Communication

In emergency situations, when conventional communication infrastructure may be disrupted, Ham radio becomes a crucial tool for connecting with emergency services, coordinating relief efforts, and seeking assistance. During natural disasters or large-scale emergencies, Ham radio operators can establish ad-hoc networks, providing a lifeline of communication when all other means fail.

Community and Knowledge Sharing

The Ham radio community promotes a culture of knowledge sharing and continuous learning. There are a considerable number of growing communities and Ham radio operators often get together and participate in training sessions, seminars and events to improve their skills and proficiency with their radio devices. This shared knowledge creates a vast network of experienced operators who can provide guidance and support during emergencies and challenging outdoor situations.

Technical Exploration and Experimentation

Amateur radio licenses provide operators with opportunities for technical exploration and experimentation. Ham radio enthusiasts can construct their own antennas, experiment with various modes of communication, and even build and modify their equipment. This provides a versatile and flexible option for radio enthusiasts. The continuous tinkering with devices and equipment increases their understanding of radio technology and also makes them able to quickly adapt and improvise during real-life scenarios.

Weather Monitoring and Reporting

Ham radio operators often participate in weather monitoring and reporting activities. They can relay vital weather information to local authorities, assisting in early warning systems and providing valuable data for weather forecasting. This contribution to public safety further solidifies the role of Ham radio operators as critical assets in disaster preparedness and response.

International Connectivity and Cultural Exchange

Ham radio enables operators to communicate with people across the globe, fostering international friendships and cultural exchange. This global reach enhances the sense of community among operators and offers unique opportunities to learn about different cultures and practices. Additionally, in emergency situations with global impacts, Ham radio operators can establish international communication links, facilitating cooperation and aid distribution.

In summary, Ham radio offers a wide array of advantages for outdoor enthusiasts and preppers. Its independence from traditional infrastructure, long communication range, and emergency capabilities make it an indispensable tool in remote and challenging environments. Moreover, the spirit of knowledge sharing and technical exploration within the Ham radio community fosters a sense of camaraderie and preparedness among operators. Whether it be for adventure, emergency, or cultural exchange, Ham radio provides a reliable and versatile means of communication that continues to thrive in the modern world.

Understanding the Value of Baofeng Radio in Survival Situations

In survival situations, where communication infrastructure may be compromised or unavailable, the value of Baofeng radios becomes truly evident. Baofeng radios are compact in size, durable, and very versatile. These features make them ideal companions for outdoor enthusiasts, preppers, and individuals seeking to be prepared for unforeseen circumstances.

One of the primary advantages of Baofeng radios in survival situations is their ability to operate in both licensed and unlicensed frequency bands. With the appropriate radio license, Ham radio operators can access a wide range of frequencies, including the VHF and UHF bands. This gives one the flexibility to communicate over long distances, and even longer when using repeater stations or favorable line-of-sight conditions. On the other hand, in unlicensed frequency bands, Baofeng radios can still provide short-range communication. As much as we have covered regulations and compliance just now, the ability to operate in unlicensed frequency bands is paramount in certain situations as it enables one to coordinate among group members during emergency situations. By establishing an emergency communication plan and having designated frequencies programmed into the radios, users can swiftly and efficiently reach out for help or coordinate rescue efforts in times of distress.

Additionally, Baofeng radios are equipped with multiple power settings, allowing users to conserve battery life when needed or maximize their transmission range in critical situations. The radios can be powered using rechargeable batteries, standard AA batteries, or even through solar chargers, making them resilient to power outages and prolonged survival scenarios.

Chapter 2: Radio Etiquette and Communication Rules

Two-Way Radio Protocols

Effective communication is the cornerstone of successful radio operations, especially in critical situations and outdoor environments. Two-way radio protocols, also known as radio procedures or etiquette, establish a standardized set of rules and practices for transmitting and receiving messages. These protocols ensure clear and efficient communication, minimize misunderstandings, and promote discipline on the airwaves. In this section, we will explore essential radio etiquette and two-way radio protocols.

Radio Etiquette

Baofeng user etiquette, like radio etiquette in general, involves a set of guidelines and best practices to ensure effective communication, minimize interference, and maintain polite interactions with other radio users.

Here are some essential Baofeng user etiquette tips:

6. **Identify Yourself:** When initiating a communication, always identify yourself and your call sign (if you are an amateur radio operator) at the beginning of the transmission. This helps others know who they are talking to and is particularly important in amateur radio communications

7. **Listen First:** Before transmitting, take a moment to listen to ongoing conversations on the frequency to avoid interrupting ongoing communications. This practice is commonly known as "listening before speaking" or "making a call."

8. **Clear and Concise Communication:** Keep your transmissions brief, clear, and to the point. Avoid lengthy conversations or unnecessary chit-chat. Be mindful of other users who may want to use the frequency.

9. **Use Plain Language:** Use standard and straightforward language to ensure easy understanding by all users. Avoid using technical terms that may not be familiar to others.

10. **Observe Channel Usage:** Different frequencies or channels may have specific purposes, such as emergency communications, public service events, or general chit-chat. Use the appropriate channels for their intended purposes.

11. **Respect Privacy:** Avoid sharing personal or sensitive information over the airwaves. Be cautious about transmitting personal details or sensitive data.

12. **Avoid Interference:** Ensure your Baofeng radio is set to the proper frequency and power level, and refrain from using frequencies allocated to other services. Unauthorized use of frequencies can cause interference and disrupt critical communications.

13. **Practice Good Radio Discipline:** Wait for a pause in ongoing conversations before transmitting (no "doubling"), avoid transmitting over someone else's transmission, and give priority to emergency or important communications.

14. **Be Patient and Polite:** If you encounter interference or unclear transmissions, be patient and allow others to complete their communications. Treat other users with respect and politeness.

15. **Emergency Communications:** If you hear distress calls or emergency communications, do not interrupt unless you have valuable information to contribute. Allow the designated emergency communications to proceed without disruption.

16. **Battery Management:** Ensure your Baofeng radio's battery is adequately charged before use. If you are using the radio for extended periods, carry spare batteries or a reliable power source.

17. **Follow Regulations:** Comply with all relevant radio regulations and licensing requirements in your country. Ensure you are authorized to use the frequencies and power levels you are using.

Standard Operating Procedures

Effective communication is paramount in any field, and when it comes to radio communication, clarity and precision are of utmost importance. Understanding the terminology and syntax used in radio communication is essential for smooth and efficient interactions over these devices.

Basic Radio Protocol

Adhering to these protocols or procedures ensures that your messages are received clearly and that your transmissions do not interfere with other operators. The table below outlines some standard operating procedures (SOPs):

Procedure	Description	Application
Clear Speech	Speak clearly and enunciate words to ensure your message is easily understood by other operators.	Applies to all radio communication.

Phonetic Alphabet	Use the NATO Phonetic Alphabetfor spelling out words and important details. (more details below).	Especially useful when communicating names, locations, or other critical information.
Call Signs	Always use your call sign when initiating communication and identify yourself at regular intervals.	Helps identify who the sender of the message is and establishes accountability during conversations.
Wait for Turn	Allow a short pause before transmitting to ensure you are not interrupting ongoing communications.	Helps prevent overlapping transmissions, reducing confusion on the airwaves.
Acknowledgment (ACK)	Use "Roger" or "Affirmative" to acknowledge receipt of a message and "Negative" to indicate negation.	Confirms that the message has been received and understood or signifies a negative response.
End of Transmission (EOT)	Conclude your communication with "Over" if expecting a response or "Out" if ending the conversation.	Notifies other operators that you have finished transmitting and are awaiting a reply or concluding the conversation.

Additional Procedures

In addition to basic protocols, certain radio procedures ensure efficient and organized communication, especially in group or emergency situations.

Procedure	Description	Application
Priority Communications	Use "Priority" to indicate an urgent message that takes precedence over regular communications.	Essential in emergency situations to ensure critical messages are received promptly.
Sequential Communication	In group communications, use a structured approach, such as "Round Robin" or "Direct Calling," to avoid confusion.	Ensures everyone gets a chance to participate and keeps communication organized.
Emergency Procedures	Establish specific procedures for emergency communications, including	Critical for quickly and effectively communicating during emergencies.

	designated channels and distress signals.	
Prohibited Language	Avoid the use of profanity or inappropriate language during radio transmissions.	Maintains professionalism and ensures a respectful radio environment.
Distress Calls	Follow standard distress call protocols, such as using "MAYDAY" three times, for life-threatening emergencies.	Alerts others to your emergency and indicates a need for immediate assistance.
Channel Usage Rules	Establish and adhere to channel usage rules to prevent interference and overcrowding on specific frequencies.	Helps maintain order and efficient use of available radio channels.

It's important to know that when using two-way radio communication, some words and phrases should be used at certain moments such as greeting, speaking to, or say goodbye to another party. This tip is one of the most important to know as some words commonly used in everyday speech may not always transmit clearly and be understandable over two-way radio waves. Using two-way radio lingo is like speaking in a secret code, ensuring messages are conveyed accurately, and reducing misunderstandings or interference.

Radio Phonetic Alphabet

The Military Alphabet, often referred to as the NATO Phonetic Alphabet, stands as an internationally recognized spelling mechanism that aids in the precise articulation of words and numerical figures. Designed to eliminate potential ambiguity, this alphabet is particularly useful in situations where letters and numbers bear phonetic similarities that could lead to confusion. Its widespread adoption is a testament to its effectiveness in enhancing clarity and understanding in communication.

A – Alpha	J – Juliet	S – Sierra
B – Bravo	K – Kilo	T – Tango
C – Charlie	L – Lima	U – Uniform
D – Delta	M – Mike	V – Victor
E – Echo	N – November	W – Whiskey

F – Foxtrot	O – Oscar	X – X-ray
G – Golf	P – Papa	Y – Yankee
H – Hotel	Q – Quebec	Z – Zulu
I – India	R – Romeo	

Radio prowords:

In the realm of radio communication, time is of the essence, and clarity is crucial. Prowords, short for "procedure words," are standardized terms used to replace longer phrases and sentences, streamlining communication and reducing the risk of misinterpretation. By adopting prowords, users of Baofeng radios can ensure efficient and effective exchanges in various scenarios. Let's explore some common radio prowords and their meanings:

Proword	Meaning	Example Use
Affirmative	Yes	"Do you copy?" - "Affirmative."
Negative	No	"Are you ready?" - "Negative, standby."
Roger	Message received and understood	"Please proceed to the designated area." - "Roger that."
Over	End of transmission, awaiting response	"Can you provide the status update?" - "Over."
Out	End of conversation	"Thank you for the information. Out."
Say again	Repeat your last transmission	"Repeat your last message; I couldn't copy."
I say again	Repeating for clarity	"Meet at 0600 hours, I say again, 0600 hours."
Break	Urgent interruption	"Break, break, emergency situation at coordinates XYZ."
Wilco	Will comply	"Please proceed to checkpoint Charlie." - "Wilco."
Radio check	Request for signal strength check	"This is Bravo team, radio check, over."
Out and	Ending transmission and leaving the	"Task complete, out and clear."

clear	radio on	
Standby	Wait and standby for further instructions	"Standby for clearance to enter restricted area."
Read back	Request to repeat a message for verification	"Please read back the coordinates I provided."
Correction	Correction to an error in the previous transmission	"Correction, the correct frequency is 146.540 MHz."
Wait out	Temporary pause in the conversation	"Wait out while I consult with my team."

Using radio prowords not only ensures efficient communication but also helps establish a professional and standardized language across diverse radio operators. These prowords have been developed and refined over time to facilitate clear and concise exchanges, especially in high-stress or emergency situations.

While prowords are valuable tools in radio communication, as civilians it is essential to strike a balance and not overuse them. Over-reliance on prowords can lead to unnecessary clutter and delay in conveying critical information. Operators must be familiar with these prowords and incorporate them judiciously into their conversations. This of course does not apply to military personal that need to follow a stricter regime.

Military Radio Communications

In military operations, effective radio communication is vital for the successful coordination and execution of missions. Across various branches of the U.S. military, including the Navy, Coast Guard, Air Force, Marines, and Army, certain communication techniques and protocols remain consistent to ensure standardized and secure exchanges. This section provides an overview of the key similarities in military radio communication, the protocols followed, and the use of specific lingo and prowords.

Key Similarities in Military Radio Communication

- Military personnel across branches employ a range of electromagnetic waves, including AM, FM, high frequency (HF), and ultra-high frequency (UHF), for transmitting messages over radio systems.
- The use of International Morse Code persists as a standard for basic communication, primarily involving a radio transmitter with an oscillator.

- To achieve precise timing for encrypted radio transmissions, the U.S. military relies on Zulu Time, also known as Coordinated Universal Time (UTC).
- The Military Alphabet, also known as the NATO Phonetic Alphabet (as depicted above), is universally utilized by military personnel to spell out call signs and messages accurately, ensuring clarity and avoiding confusion.
- Uniformity in radio lingo ensures consistent communication between different branches of the military, facilitating interoperability and seamless cooperation.

Specific Military Radio Communication Protocols

Two-way radio communication in the military adheres to a set of universal rules, but certain protocols are more restrictive due to the sensitive nature of military operations and the imperative to safeguard national interests. Best practices include:

- **Identify with Call Signs:** Communicate with designated call signs to establish clear connections and protect identities.

- **Brief Pause After PTT Press:** After pressing the "push-to-talk" (PTT) button, pause momentarily to avoid cutting off the initial part of the message.

- **Be Direct and Concise:** Keep messages direct and brief to ensure swift and efficient communication.

- **Speak Slowly and Clearly:** Enunciate clearly and speak at a moderate pace to enhance message comprehension.

- **Use Military Alphabet:** Employ the Military Alphabet when spelling out letters and numbers to minimize errors in transmission.

- **Leverage Lingo and Prowords:** Use appropriate lingo and prowords to reduce ambiguity, shorten messages, and maintain a standardized language.

Common Military Prowords

Apart from the prowords described previously, there are specific military prowords that have particular meanings:

Proword	Meaning
ACKNOWLEDGE	Directive requiring the recipient to confirm they have received the message.

ALL AFTER	This refers to a part of the message as "all that follows."
ALL BEFORE	This refers to a part of the message as "all that proceeds."
AUTHENTICATE	This is used by sender to request the called station to authenticate the message that follows.
AUTHENTICATION IS	Specifies the authentication of the transmitted message.
BREAK	Used to introduce a pause before relaying the next part of the message.
CLEAR	Clear the channel to relay a message of higher importance.
CORRECT	Confirms the message broadcasted is correct.
CORRECTION	Corrects a previously misheard message.
DISREGARD THIS TRANSMISSION-OUT	Indicates an error and that the message should be disregarded.
DO NOT ANSWER	Indicates no response is expected. Sender ends with "OUT."
EXEMPT	Designates recipients who should disregard the message.
FIGURES	Indicates that numbers will follow in the transmission.
FROM	Identifies the originator of the message.
GROUPS	Signifies that the message contains numbers of groups.
I AUTHENTICATE	Sender is authenticating the following message.
IMMEDIATE	Used for urgent situations requiring immediate attention.
INFO	Requests information from addressees immediately.
I READ BACK	Instructs recipient to repeat the instructions back for confirmation.
I SAY AGAIN	Repeats a message due to misunderstanding or importance.
I SPELL	Spells out following words phonetically using the NATO Phonetic Alphabet.
I VERIFY	Used to verify a request and repeated to verify a sent message.
MESSAGE	Indicates that a message must be recorded.
MORE TO FOLLOW	Indicates more information will be transmitted.

OUT	Ends the transmission.
OVER	Ends message and awaits a reply.
PRIORITY	Indicates a message of high importance, taking precedence.
READ BACK	Requests recipient to repeat the message for confirmation.
RELAY (TO)	Instructs transmitting the message to specified call sign(s).
ROGER	Confirms receipt of a message.
ROUTINE	Conveys a routine and less important message.
SAY AGAIN	This is used to ask a sender to repeat their last transmission.
SILENCE	Instructs an immediate stop of communication and be silent
SILENCE LIFTED	Lifts a temporary communication silence.
SPEAK SLOWER	Requests slower speech.
THIS IS	Transmits a message from one call sign to another. e.g. "Echo 1, Echo 2, over" vs. "Echo 1 THIS IS Echo 2, over."
TIME	Specifies the time frame for complying with the message.
TO	Addresses those who must comply with the message.
UNKNOWN STATION	Indicates an unknown station attempting communication.
VERIFY	Requests verification of a message.
WAIT	Instructs sender or recipient to pause for a few seconds.
WILCO	Indicates receipt and compliance with instructions.
WORD AFTER	The word of the message to which I have reference is that which follows … ____.
WORD BEFORE	The word of the message to which I have reference is that which proceeds … ____.
WORD TWICE	Repeats a word for improved understanding.
WRONG	This is used to say your last transmission was incorrect. The correct version is ____.

Example of Military Radio Communication

To illustrate the practical application of military radio communication, consider the following example between two military personnel:

[Eagle 1:] "Eagle 2, Eagle 1, over."

[Eagle 2:] "Go ahead Eagle 1, over."

[Eagle 1:] "Eagle 2, enemy spotted 2 kilometers west, break ..."

[Eagle 1:] "Take cover 2 kilometers east at Alpha Shack, read back, over."

[Eagle 2:] "I read back: enemy spotted 2 kilometers west. Take cover 2 kilometers east at Alpha Shack, over."

[Eagle 1:] "Correct, over."

[Eagle 2:] "Wilco, over."

[Eagle 1:] "Roger, out."

N.B. The senior-ranking official typically initiates conversations and concludes with "OUT" as a sign of respect and protocol.

Call Signs

Call signs can be letter and number combinations or one or more pronounceable words used mainly to establish communication and provide a unique identifier for a transmitter station or individual. They are essential in military radio communication for identifying communication activities while safeguarding individuals' identities. These designations change frequently to maintain anonymity and prevent breaches in security protocols.

Chapter 3: Baofeng Radios: Models and Features

Introduction to Popular Baofeng Radio Models

There are several Baofeng radios models available in the market. Each one is designed to cater to specific needs and user preferences.

Below is a list of the most famous models:

The **Baofeng UV-5R** is the most widely known model from this company. I have already discussed it in the Introduction to this handbook, but to recap it here, this is a highly versatile model known for its compact design and user-friendly interface. Its dual-band functionality allows users to communicate on both Very High Frequency (VHF) and Ultra High Frequency (UHF) bands. The UV-5R operates in various power settings, so that users can adjust the output power to save on battery life or even extend communication range as needed. Its LCD display makes essential information available at a glance, including the current channel, frequency, and battery status. This is also one of the most affordable models.

The **Baofeng BF-F8HP** is well known for its high power output, and it is appreciated for extended communication range. This model features a high-gain antenna, further enhancing signal reception and transmission, making it an excellent choice in case of emergency and it is generally recommended for outdoor activities in remote areas. The BF-F8HP also boasts improved audio clarity, securing clearer communication even in noisy environments.

The **Baofeng UV-82HP** is a more robust known for its tactical design and enhanced performance. It is considered one of the most reliable models. Its tri-color display allows users to customize the backlight color for specific functions, facilitating the identification of different features. With its dual push-to-talk (PTT) buttons, users can effortlessly switch between channels, making communications easier and effective. The UV-82HP operates on both VHF and UHF bands and its sturdy construction and extended battery life make it well-suited for extended outdoor excursions and challenging environments.

The **Baofeng RD-5R**'s main peculiarity is that it supports both digital and analog communication modes. As a dual-band radio, it provides users with the possibility to choose between traditional analog communication and the digital mode which provides privacy and better audio quality. Thanks to its features this model is valued for the seamless communication with users on both digital and analog radio systems.

The **Baofeng UV-5R** stands as one of the most widely recognized handheld dual-band two-way radios in the market. With its dual-band capability, the UV-5R can operate on both VHF and UHF frequencies. It's equipped with an LCD display and a numeric keypad which allows for easy manual input of frequencies and navigation through its menu options. As an extra, its FM radio feature adds a touch of entertainment, enabling users to listen to regular FM broadcast stations when not engaged in two-way communications. The UV-5R's support for VOX and dual watch further adds to its appeal, making hands-free operation possible and allowing monitoring of two channels simultaneously.

The **Baofeng BF-F8HP** serves as an upgraded version of its predecessor the UV-5R. The BF-F8HP boasts a tri-power feature, offering selectable power output levels of 1 watt, 4 watts, and 8 watts. This additional power setting allows for an extended communication range. With an upgraded antenna included in the package, the BF-F8HP ensures better performance and improved signal transmission compared to the standard UV-5R antenna. The higher-capacity 2100mAh battery makes this model stand out allowing extended operating times on a single charge.

The **Baofeng UV-82HP** offers a blend of features that combine the best aspects of both the UV-5R and the BF-F8HP. Like the BF-F8HP, the UV-82HP also includes the tri-power capability, presenting users with the option to switch between 1 watt, 4 watts, and 7 watts of power output. This flexibility in power levels allows for better adaptability to diverse communication scenarios. The UV-82HP comes equipped with dual push-to-talk (PTT) buttons, a unique feature enabling users to transmit on different frequencies without the need to switch channels. This added feature makes this model appealing to those users seeking seamless communication in dynamic environments. With a rugged design, the UV-82HP is built to withstand more demanding conditions, making it suitable for outdoor activities and harsher settings. Its durable construction, combined with its tri-power functionality and dual PTT buttons, has garnered favor among those seeking a robust and feature-rich handheld radio.

Understanding Key Features and Functions

The majority of Baofeng radios have several key features and functions in common. Being familiar with these will allow you to make informed decisions when choosing which model works best for your needs and circumstances.

Let's explore what the key features and functions are:

Most Baofeng radios support **dual-band operation**, meaning they work on both Very High Frequency (VHF) and Ultra High Frequency (UHF) bands. This dual-band capability allows users to access a wide range of frequencies, offering greater flexibility and communication options.

Baofeng radios typically cover a broad **frequency range**, spanning from approximately 136 to 174 MHz for VHF and 400 to 520 MHz for UHF. This extended frequency coverage enables communication across various channels and bands.

Most Baofeng radios feature a clear and informative **LCD display**, showing things like channel number, frequency, signal strength, battery status, and more, making the user experience easier than with other radios.

Baofeng radios support **Continuous Tone-Coded Squelch System (CTCSS) and Digital-Coded Squelch (DCS) encryption**. These privacy features allow users to communicate on specific sub-audible tones or digital codes, minimizing interference from other radio users on the same frequency.

The majority of Baofeng models use **rechargeable lithium-ion batteries**, offering reliable power for extended usage. Some of them even include options for using standard AA batteries or battery packs, providing backup power sources for emergencies.

Voice-Operated Exchange (VOX) is another feature common to many Baofeng radios. This enables hands-free communication and when activated, allows the radio to detect the user's voice and automatically transmit without the need to press the Push-to-Talk (PTT) button.

Almost all Baofeng radios come equipped with **channel memory**, allowing users to save and access their preferred frequencies and settings. This feature simplifies communication, especially when switching between frequently used channels.

Many Baofeng radios have a **Push-to-Talk (PTT)** button located on the side of the device, facilitating quick and easy transmission. Additionally, they often feature programmable side keys that can be customized for specific functions, enhancing operational efficiency.

Baofeng radios offer **adjustable power output settings**, allowing users to conserve battery life when communicating over shorter distances or boost power for long-range communication. This flexibility is especially valuable during emergencies, where conserving battery power is crucial.

All Baofeng radios are compatible with a wide range of **accessories** - external antennas, speaker-microphones, and programming cables, allowing you to further customize your radio according to your needs.

In order to prevent accidental changes to settings while on the move, Baofeng radios often offer a **key lock functionality**, which once activated disables the radio's buttons, ensruting that critical configurations cannot be changed.

Baofeng radios are equipped with an audio **output jack**, so that users can connect external audio accessories, such as earpieces or headsets, allowing discrete communication and better noise control.

Baofeng radios also typically incorporate **channel scanning** functionality, where users can automatically scan through preset or user-defined channels. This feature allows them to quickly locate active channels and monitor communications in their vicinity.

Not all but many Baofeng models come equipped with a **built-in flashlight**, which can prove especially useful during emergencies.

Speaking of emergencies, certain Baofeng radios come with an **emergency SOS alarm** feature. When activated, the radio emits a loud distress signal to alert nearby users encouraging a rapid response and assistance.

Finally, Baofeng radios with **data transmission capabilities** include the possibility for users to send and receive simple data messages.

Not all models include all the function listed above, so be careful when purchasing your radio and make sure you have done your research to avoid finding yourself out there with a model that doesn't meet your needs and requirements.

Choosing the Right Baofeng Model for Your Needs

Choosing the right Baofeng model for your needs involves a more detailed assessment of technical specifications, features, and considerations based on your specific use case. Below I have listed some of the things you need to consider when approaching the final decision, but ultimately you are the only one who knows what you need.

Frequency Bands and Dual-Band Capability

If you're primarily interested in amateur radio communication, dual-band radios like the Baofeng UV-5R are popular choice. These radios can operate on both Very High Frequency (VHF) and Ultra High Frequency (UHF) bands, giving you access to a wider range of frequencies, repeaters, and communication options. VHF frequencies are commonly used for local and regional communications, while UHF frequencies offer better penetration through obstacles and buildings, making them suitable for more urban environments. However, if you have specific use cases that require only one band, you can opt for a single-band model, which might be more straightforward to operate and potentially even more affordable.

Power Output and Communication Range

The power output of a Baofeng radio significantly influences its communication range. Baofeng radios like the UV-5R and BF-F8HP offer selectable power levels, usually ranging from 1 watt (low power) to 4 or 8 watts (high power). Higher power output generally results in extended communication range, especially in open areas with line-of-sight communication paths. If you plan to use the radio for emergency communications, outdoor adventures, or in areas with limited repeater coverage, a higher-power model like the BF-F8HP may be a better option. However, also consider what is your required battery life span as higher power levels means more energy consumption and a shorter battery autonomy.

Features and Functionality

Consider the specific features offered by each Baofeng model and how they align with your needs. For instance, if you enjoy listening to FM radio stations during downtime, the UV-5R's FM radio receiver might be a valuable addition. If you require hands-free operation, models with Voice-

Activated Transmission (VOX) can be beneficial. Some Baofeng radios support dual watch, enabling you to monitor two channels simultaneously, which can be valuable for staying updated on multiple frequencies without having to switch constantly.

Build Quality and Durability

Depending on your intended use, you might prioritize the build quality and durability of the radio itself. If you plan to use the radio for outdoor activities like hiking, camping, or fieldwork, models with a rugged design like the UV-82HP might be more suitable. These radios often come with reinforced casings, robust antenna connections, and a sturdier feel. On the other hand, if you plan to primarily use the radio indoors or in less demanding conditions, the UV-5R's compact and lightweight design might suffice.

Licensing and Compliance

Ensure that the Baofeng model you select aligns with the radio regulations and licensing requirements in your area of operation. In many regions, amateur radio operators need to obtain a license to operate on specific frequencies and power levels. The UV-5R and other Baofeng radios provide access to a wide range of frequencies, including amateur radio bands, but using them without the proper authorization can lead to legal consequences. Always be aware of what regulations are in vigor in your area and what they involve.

Price and Budget

Budget is an essential factor when selecting a Baofeng model. While the UV-5R is often recommended for its affordability, other models may offer additional features or enhanced capabilities at slightly higher price points. You have to determine whether the features offered by higher-end models, such as the BF-F8HP or UV-82HP, justify the additional cost for your specific use case.

Case Study: Choosing the Right Baofeng Model for Outdoor Adventures and Checklist

Introduction

Meet John, an outdoor enthusiast and amateur radio operator, who loves to explore nature and engage in a number of outdoor activities such as hiking, camping, and off-road adventures, you name it. John recently decided to invest in a Baofeng radio to enhance communication during his outdoor excursions and potentially engage in amateur radio operations. However, with multiple Baofeng models available, he finds himself overwhelmed by the options and struggles to determine the best fit for his needs. To make an informed decision, John decides to conduct a thorough assessment of each model based on his specific requirements.

Objectives

John's main objectives for choosing a Baofeng model are:

1. **Reliable Communication:** John needs a radio that offers reliable communication over various terrains, including forests, mountains, and remote areas.

2. **Extended Range:** As John often take off towards challenging trails and off-road adventures, he needs a radio with good transmission range to ensure seamless communication with his fellow adventurers.

3. **Rugged Design:** John has never been the careful type and since John's outdoor activities involve exposure to rough environments and weather conditions, he seeks a radio that can survive the rigors of outdoor adventures.

4. **Dual-Band Functionality:** As an amateur radio operator, John also wants access to a wide range of frequencies for ham radio communications and the ability to connect with other operators locally and globally.

5. **Budget-Friendly:** While John wants a radio with enhanced features, he also aims to stay within a reasonable budget.

Evaluation Process

John recognizes the importance of dual-band capability for accessing multiple frequencies. After reviewing the specifications, he narrows down his options to the Baofeng UV-5R and the Baofeng BF-F8HP, both of which offer dual-band functionality. The UV-5R's affordability appeals to him, but he acknowledges that the BF-F8HP offers higher power output and extended communication range, making it more suitable for his adventurous outings.

Then, considering his need for extended range, John closely examines the power output of each radio. The BF-F8HP stands out with its tri-power capability, offering 1 watt, 4 watts, and 8 watts power options. The UV-5R, on the other hand, has lower power settings, but John values the portability and affordability of this model. After weighing the pros and cons, he leans toward the BF-F8HP for its potential to provide better range and communication performance.

While inclined towards the BF-F8HP John still assess the features of both models. The UV-5R includes an FM radio receiver which is nice to have, but John realistically acknowledges that he rarely uses this feature during outdoor adventures. On the other hand, the BF-F8HP's VOX capability and dual watch function are appealing, as they allow for hands-free communication and monitoring of two channels simultaneously, facilitating better situational awareness. These features align well with John's desire for ease of use and efficient communication.

Given the rough outdoor conditions he faces, John also prioritizes a rugged design. The UV-5R's compact build is suitable for casual use, but he decides that the UV-82HP, with its robust construction and sturdier feel, better meets his requirements. The UV-82HP's durability ensures it can withstand the challenges of his adventurous lifestyle.

Finally, as a licensed amateur radio operator, John verifies that both the UV-5R and BF-F8HP offer access to the ham radio bands and comply with relevant radio regulations. He is pleased that both models align with his licensing requirements, allowing him to engage in amateur radio activities legally.

Final Decision

After thorough evaluation and consideration, John concludes that the Baofeng BF-F8HP is the best fit for his needs. The tri-power capability and rugged design offer the extended range and durability required for his outdoor adventures. The VOX and dual watch features enhance his communication experience, allowing for hands-free operation and effective monitoring of multiple channels. While the UV-5R's affordability is attractive, the additional capabilities of the BF-F8HP make it worth the investment for John's outdoor activities and amateur radio pursuits. With his decision made, John confidently purchases the Baofeng BF-F8HP and looks forward to seamless communication and exciting adventures in the great outdoors.

The above is just a simple example, but it shows the train of thoughts you are likely to go through when you get to the stage of choosing your radio. To make things smoother, I have created a checklist for you to follow and understand which Baofeng model suits you most.

By following this checklist, you can systematically evaluate the technical specifications, features, and considerations of each Baofeng model, helping you make an informed decision that aligns with your communication needs and intended use. Remember to prioritize the factors that matter most to you, such as frequency bands, power output, features, and budget, to find the ideal Baofeng radio that suits your requirements.

Baofeng Radio Model Checklist

1. **Frequency Bands and Dual-Band Capability:**
   - Determine if you need a single-band or dual-band radio.
   - Consider the frequencies you need to access and whether a dual-band model like the Baofeng UV-5R provides the versatility you require.

2. **Power Output and Communication Range:**
   - Assess the power output options of each model (e.g., 1 watt, 4 watts, 8 watts).
   - Determine if you need extended communication range for your intended use, such as outdoor adventures or amateur radio operations.

3. **Features and Functionality:**
   - Identify essential features, such as VOX, dual watch, FM radio receiver, and more.
   - Choose a model that aligns with your need for hands-free operation, monitoring multiple channels, or other specific functionalities.

4. **Build Quality and Durability:**

- Consider your intended use and whether a rugged design, like the Baofeng UV-82HP, is necessary for outdoor activities and challenging environments.
- Assess the build quality and sturdiness of each model to ensure it can withstand your expected conditions.

5. **Licensing and Compliance:**
   - If you're an amateur radio operator, ensure the model provides access to the ham radio bands and complies with relevant radio regulations.
   - Verify that your chosen model meets the licensing requirements in your country or region.

6. **Budget and Affordability:**
   - Determine your budget and weigh the features and capabilities of each model against its cost.
   - Consider the affordability and popularity of models like the Baofeng UV-5R, while also exploring higher-end models like the Baofeng BF-F8HP or UV-82HP for potential additional benefits.

7. **Reviews and User Feedback:**
   - Read reviews and user feedback for each Baofeng model you're considering.
   - Pay attention to real-world experiences and testimonials to understand how each radio performs in various scenarios.

8. **Aftermarket Support and Accessories:**
   - Check the availability of aftermarket support, such as compatible accessories (e.g., antennas, batteries, programming cables).
   - Consider the availability of resources, tutorials, and online communities for your chosen model.

What Can You Do with Your Baofeng Radio?

Baofeng radios, such as the popular UV-5R, have become a well known product in the world of communication. Thanks to their dual-band capability, licensed amateur radio operators can explore many different frequencies and bands, engaging in local and global conversations, experimenting with various modes, and participating in digital data transmission. While many uses them to cultivate their hobby, these radios have a much bigger role when it comes to emergency communication and preparedness, having proven to be instrumental during times of natural disasters and crises when traditional infrastructures might fail. Amateur radio operators, as part of organizations like ARES and RACES, play critical roles in providing essential communication support to emergency responders and communities in need.

For outdoor enthusiasts, Baofeng radios offer an extra layer of security when practicing adventure sports and other more or less extreme outdoor activities. Hikers, campers, off-road enthusiasts, and wilderness explorers rely on these radios for group coordination, sharing information, and ensuring safety in remote areas. The rugged design of certain Baofeng models ensures durability in harsh environments, withstanding dust, water, and impacts encountered during outdoor pursuits.

Beyond amateur and outdoor applications, Baofeng radios also find practical uses in the commercial sector. Industries like security, event management, construction, and hospitality benefit from the cost-effectiveness and versatility of these radios. They enable efficient communication between team members within various sectors. Like with amateurs, businesses too have to adhere to regulatory requirements and obtain proper licenses for commercial radio use, ensuring responsible and lawful communication practices.

It may come as a surprise but Baofeng radios also serve as educational tools, especially for those interested in radio technology and wireless communication. Aspiring radio operators and electronics enthusiasts can gain hands-on experience with programming frequencies, experimenting with different antennas, and understanding radio propagation principles.

Chapter 4: Programming Your Baofeng Radio

Step-by-Step Guide to Programming Frequencies and Channels

The table below is an easy-to-follow guide to programming frequencies and channels on your Baofeng radio. Familiarize yourself with these steps which willhelp you customize your radio settings in line with your needs.

Step	Procedure
1 Power on the radio	Turn on the radio by rotating the "Power/VOL" knob clockwise
2 Input desired frequency	Use the keypad to input the desired frequency, including the decimal point (e.g., 146.520 MHz for the 2-meter band)
3 Choose the "SAVE" option	Choose the frequency and press the "MENU" button, then use the arrow keys to navigate to the "SAVE" option
4 Confirm frequency setting	Confirm the frequency setting by pressing the "MENU" button again
5 Select memory channel	Access the memory channel where you want to store the programmed frequency
6 Store frequency on memory channel	Press the "MENU" button, then navigate to the "SAVE" option to store the frequency on the selected memory channel
7 Confirm "SAVE action"	Confirm the save action by pressing the "MENU" button again
8 Enable transmission	To transmit on the programmed frequency, ensure the "TX" indicator is displayed on the screen
9 Adjust transmission power	Use the arrow keys to adjust the transmission power (high or low) according to your needs
10 Save transmission power settings	Save the transmission power setting by pressing the "MENU" button and selecting the "SAVE" option
11 Enable reception	To receive signals, ensure the "RX" indicator is displayed on the screen
12 Adjust volume	Adjust the volume using the "Power/VOL" knob to a comfortable level
13 Set up dual-frequency channel	To set up a dual-frequency channel, repeat steps 2 to 12 for the second frequency
14 Toggle between dual frequencies	Use the "A/B" button to toggle between the two programmed frequencies on the dual-frequency channel
15 Lock the keypad	If needed, lock the keypad to prevent accidental changes to your programmed frequencies
16 Adjust squelch levels	Adjust squelch levels to eliminate unwanted background noise when no signal is present

17 Enable power-saving features	Enable power-saving features, such as "Battery Saver" and "VOX," to conserve battery life
18 Monitor dual standby	Monitor two frequencies simultaneously with the "Dual Standby" feature for increased convenience
19 Check for updates	Stay up-to-date with firmware and software updates for enhanced performance and functionality
20 Adjust frequency range	Utilize the "Frequency Range" option to adjust the radio's frequency coverage based on your specific needs and regulatory requirements

These steps will help you program frequencies and channels on your Baofeng radio. Let's look into each step in more details.

Programming with CHIRP Software for Ease and Efficiency

CHIRP is an open-source programming software designed for a number of radio models, including Baofeng radios. It offers a wide range of features that make the programming process much simpler.

To get started, download and install the CHIRP software on your computer. Once installed, use a cable to connect your Baofeng radio to your computer. Upon launching CHIRP, you'll see a number of radio settings and configuration options for you to choose from. The software also allows you to read and download the existing settings from your radio, including the possibility to do a backup of your existing configurations. This way you can always revert to your original settings if needed.

Programming your Baofeng radio using CHIRP is fairly straightforward. You can manage all aspects of your radio's programming - such as frequencies, memory channels, power settings, squelch levels, and more from the same dashboard.

To make any changes to your setting simply select the change and save. Once your configurations are set, upload them back to your Baofeng radio. This is a two-way data transfer between your computer and radio, that drastically simplifies the whole programming process.

In addition to what discussed, CHIRP also provides additional features, such as the ability to copy and paste configurations between different radios. This possibility is specially useful to those users owning multiple Baofeng radios as it allows them to save a lot of time. You can even import and export settings to share your configurations with others or save them as backups.

CHIRP is supported by an active community of users and developers who contributes to its ongoing improvement, ensuring the software is updated with the latest radio models and features.

You don't have to use CHIRP if you don't want to, but this is surely a recommended tool.

Customizing Settings for Optimal Performance

Customization is a mandatory step in your Baofeng radio set up process. This is the only way to make sure your radio is set up in a way that is optimal for the use you intend for it and your communication requirements.

Below are some of the key settings and features that can be customized to mold your radio's performance to your needs.

Transmit Power

Baofeng radios normally offer adjustable transmit power levels, allowing you to set the balance between saving battery and extending communication range. As a rule, for short-range communications within a group, lower power settings can be sufficient, while for higher power may be necessary for long-range or emergency situations.

How to:

- Press the "Menu" button to enter the menu mode.
- Use the arrow keys to navigate to the "Transmit Power" option.
- Select the desired power level (e.g., Low, Medium, High) using the arrow keys.
- Press the "Menu" button again to save the setting.

Squelch Level

Squelch is a noise filter that mutes the audio output of your radio when no signal is present. Customizing the squelch level blocks background noise and interference, improving the clarity of received transmissions.

How to:

- Enter the menu mode by pressing the "Menu" button.
- Navigate to the "Squelch Level" option using the arrow keys.
- Adjust the squelch level using the arrow keys to eliminate background noise.
- Save the setting by pressing the "Menu" button.

VOX (Voice-Operated Exchange)

The VOX feature enables hands-free operation by automatically initiating transmission when the radio detects your voice. Adjusting the VOX sensitivity allows you to control the threshold at which the radio activates transmission.

How to:

- Access the menu mode and find the "VOX" option.
- Use the arrow keys to adjust the VOX sensitivity (higher sensitivity for lower ambient noise).
- Press the "Menu" button to confirm and save the setting.

Scan Functions

Baofeng radios often have scanning capabilities which allow you to monitor multiple channels for activity. Customizing scan lists and priorities ensures you focus on essential frequencies and avoid missing critical communications.

How to:

- Press the "Scan" button to enter scan mode.
- Use the arrow keys to add or remove channels from the scan list.
- Adjust the scan priority by selecting the desired channel to monitor first.
- Press the "Scan" button again to exit scan mode and save the settings.

CTCSS/DCS Tones

Continuous Tone-Coded Squelch System (CTCSS) and Digital-Coded Squelch (DCS) tones prevent your radio from receiving unwanted transmissions from other users set on the same frequency. Programming the appropriate CTCSS/DCS tones helps you secure private and interference-free communications.

How to:

- Access the menu mode and locate the "CTCSS" or "DCS" option.
- Use the arrow keys to select the desired tone frequency or code.
- Save the setting by pressing the "Menu" button.

Display and Backlight Settings

Adjusting the display brightness and backlight duration helps saving battery while also ensuring visibility in various lighting conditions.

How to:

- Press the "Menu" button to enter the menu mode.
- Navigate to the "Display" or "Backlight" option.
- Adjust the brightness level and backlight duration using the arrow keys.
- Confirm and save the setting by pressing the "Menu" button.

Keypad Lock

Enabling the keypad lock prevents accidental changes to your radio's settings, ensuring that your configurations remain unchanged while in use.

How to:

- Enter the menu mode and find the "Keypad Lock" option.
- Enable or disable the keypad lock as needed.
- Press the "Menu" button to save the setting.

Timeout Timer

The timeout timer automatically stops transmission after a pre-defined period, preserving battery life.

How to:

- Access the menu mode and locate the "Timeout Timer" option.
- Set the desired timer duration using the arrow keys.
- Save the setting by pressing the "Menu" button.

Dual Watch and Dual PTT

Some Baofeng radios support dual watch. This allows you to monitor two frequencies simultaneously. By customizing dual PTT settings you can switch between primary and secondary frequencies as you wish.

How to:

- Press the "Dual Watch" or "Dual PTT" button to activate the feature.
- Use the arrow keys to switch between primary and secondary frequencies.
- Press the "Dual Watch" or "Dual PTT" button again to deactivate the feature.

Setting Alarms and Alerts

Baofeng radios offer options to set alarms for specific events, e.g., channel activity, low battery, or incoming calls, etc. Customizing these alerts allows you to pick what you get notified and ensure you don't miss important events.

How to:

- Enter the menu mode and find the "Alarms" or "Alerts" option.

- Configure the type of alarm (e.g., channel activity, low battery) and desired alert sound.
- Save the settings by pressing the "Menu" button.

Repeater Settings

Customizing the offset frequency and transmit tones ensures successful repeater access.

How to:

- Access the menu mode and locate the "Repeater" option.
- Set the desired offset frequency using the arrow keys.
- Adjust the transmit tone (if required) using the menu options.
- Press the "Menu" button to save the repeater settings.

N.B. These are generic guidelines and the instructions can vary slightly from one Baofeng model to another.

Chapter 5: Enhancing Range and Signal Quality

Best Practices for Improving Radio Range and Signal Clarity

Achieving reliable and clear communication is paramount in radio operations, especially in critical situations and outdoor environments. To enhance the range and signal clarity of Baofeng radios, operators must employ a combination of technical knowledge, optimal settings, and appropriate equipment. In this section, we will explore the best practices and techniques to maximize radio range and ensure clear transmission signals.

Antenna Selection and Tuning:

The antenna is a critical component that significantly impacts the performance of Baofeng radios. Choosing the right antenna type and length is essential for achieving the desired range and signal clarity. In following sections we will discuss further different types of antennas.

Tuning the antenna is equally important. An antenna tuner, also known as an impedance matching network, can optimize the antenna's impedance to match the radio's output. Tuning the antenna to the operating frequency is vital for maximizing communication efficiency. When an antenna is correctly tuned, it matches the radio's output impedance, minimizing signal reflections and maximizing power transfer. In some cases, the antenna may not be easily tuned to a specific frequency. In such situations, a tuner or matching network (antenna tuner) can help match the antenna's impedance to the radio's.

Also consider using an SWR meter to tune your antenna. Connect the SWR meter between the radio and the antenna and set the radio to the desired operating frequency. Adjust the antenna's length or tunable elements until you achieve the lowest SWR reading possible. A low SWR indicates that the antenna is well-matched to the radio, resulting in optimal performance.

Height and Line-of-Sight:

Radio signals travel in straight lines and can be obstructed by obstacles like buildings, hills, and dense foliage. To maximize range, elevate the radio's antenna and position yourself in a location with a clear line-of-sight to the receiving station. In outdoor settings, being on higher ground or using elevated structures like masts can significantly improve communication reach.

Consider Signal Propagation Factors:

Understanding signal propagation factors can aid in optimizing radio range and clarity. In VHF and UHF bands, radio waves are generally influenced by line-of-sight propagation. For shorter-range communications, such as within a local group, direct line-of-sight may suffice. However, in longer-distance communication, tropospheric ducting, ionospheric reflection, and ground wave propagation may play significant roles.

Transmit Power and Modulation:

Baofeng radios typically offer multiple transmit power levels. When operating in close proximity, using lower power settings can conserve battery life and reduce the risk of signal overloading. However, for long-range communication, selecting higher power levels may be necessary. Ensure compliance with local regulations and frequency band allocations when adjusting transmit power.

Using the appropriate modulation mode is crucial for signal clarity. For voice communication, Frequency Modulation (FM) is commonly employed due to its resilience against noise and interference. On the other hand, Morse code communication may utilize Continuous Wave (CW) modulation. Selecting the right modulation mode for the intended communication purpose can improve signal intelligibility.

Implement CTCSS and DCS:

Continuous Tone-Coded Squelch System (CTCSS) and Digital-Coded Squelch (DCS) are sub-audible tones and codes, respectively, used for selective calling and reducing interference. By programming CTCSS or DCS codes on Baofeng radios, users can ensure that their devices only receive signals with matching codes, minimizing unwanted interference from other users in busy radio environments.

Use External Signal Boosters:

In scenarios where extended communication range is vital, consider using external signal boosters or amplifiers. These devices can augment the radio's transmit power and enhance signal reception. However, be mindful of legal limitations and potential interference issues while employing such equipment.

Maintain and Optimize Baofeng Radio:

Regular maintenance and optimization of Baofeng radios are essential for maintaining optimal performance. Keep the radio and antenna clean and free from dirt or corrosion. Regularly check for loose connections or damaged components that may degrade signal quality. Additionally, ensure that the radio's firmware and software are up-to-date to benefit from any performance improvements or bug fixes provided by the manufacturer.

Selecting and Tuning the Right Antenna for Your Scenario

Baofeng radios support a wide range of antennas, each designed to suit different applications and operating frequencies. The choice of antenna type depends on a number of factors: the specific application, operating frequency, desired range, and environmental conditions. By selecting the appropriate antenna, radio operators can optimize their communication performance and achieve better signal clarity for the specific use case.

Here are some considerations when choosing an antenna for your scenario:

1. **Operating Frequency Range:** Ensure that the selected antenna covers the operating frequency range of your Baofeng radio. Baofeng radios typically operate in VHF (136-174MHz) and UHF (400-520MHz) bands, so make sure the antenna supports these frequencies.

2. **Gain and Direction:** Higher gain antennas, such as Yagi and ground plane antennas, provide better signal strength and directionality. Consider using high-gain antennas when you need to extend communication range in specific directions. Also consider the position of the antenna and any directionality.

3. **Portability and Size:** For portable or handheld use, compact and flexible antennas like rubber duck or telescopic antennas are convenient choices. They are lightweight and easy to carry, making them suitable for outdoor activities and on-the-go communication.

4. **Base Station Setups:** In fixed base station setups, magnetic mount antennas or ground plane antennas can be advantageous. They offer enhanced performance when mounted on metal surfaces or when better signal propagation is required.

5. **Mobile Communication:** For mobile use in vehicles, magnetic mount antennas are preferred due to their ease of installation and ability to improve signal reception while on the move.

Antenna Type	Description	Frequency Range	Application
Rubber Duck Antenna	Standard flexible rubber antennas supplied with most Baofeng handheld radios.	VHF/UHF (136-174MHz, 400-520MHz)	Suitable for short-range communication and portable use.
Whip Antenna	Longer and more rigid antennas that can be attached to the radio's SMA connector.	VHF/UHF (136-174MHz, 400-520MHz)	Provides better gain than rubber duck antennas, improving range and signal quality.

Telescopic Antenna	Extendable antennas with adjustable lengths for VHF and UHF frequency bands.	VHF/UHF (136-174MHz, 400-520MHz)	Offers variable gain based on length and allows optimization for specific frequencies and conditions.
Dual-Band Antenna	Combines VHF and UHF elements into a single antenna, allowing transmission on both bands.	VHF/UHF (136-174MHz, 400-520MHz)	Convenient for users who frequently switch between VHF and UHF frequencies.
Magnetic Mount Antenna	Designed for use with magnetic mounts, offering better performance on metal surfaces.	VHF/UHF (136-174MHz, 400-520MHz)	Ideal for mobile or base station setups where a magnetic mount can be attached to metal surfaces.
Ground Plane Antenna	Utilizes radials or a ground plane to enhance signal propagation and reduce interference.	VHF/UHF (136-174MHz, 400-520MHz)	Suitable for base station setups or when enhanced signal propagation is required.
Yagi Antenna	High-gain directional antenna with multiple elements for focused transmission and reception.	VHF/UHF (136-174MHz, 400-520MHz)	Provides excellent range and signal directionality, useful for specific point-to-point communication.
Discone Antenna	Broadband antenna designed to cover a wide frequency range, including VHF and UHF bands.	VHF/UHF (25-1300MHz)	Suitable for wideband scanning, receiving various frequencies within the VHF and UHF spectrum.
Dipole Antenna	Simple antenna with two elements, usually used for specific frequencies within the VHF/UHF bands.	VHF/UHF (136-174MHz, 400-520MHz)	Provides basic omnidirectional performance for specific frequencies and applications.

Baofeng radios utilize three main families of external antennas, catering to different use cases: portable antennas for handheld transceivers, mobile antennas for vehicle mounting, and larger antennas for fixed base stations.

Portable antennas, commonly known as "Rubber duck antennas", come as standard with Baofeng radios but are not the most efficient. Aftermarket options are available, providing significantly improved signal strength (up to eight times better performance in some cases), representing a substantial upgrade at a nominal cost, often below $20. If available, assess antennas by their gain figure, or follow the general guideline of favoring larger antennas, ensuring compatibility with your desired frequency bands. Experimenting with multiple antennas can be beneficial, with potential resale options.

Mobile antennas offer two key benefits over internal antennas in vehicles. Firstly, they are mounted outside the vehicle, avoiding partial shielding and obstructions, leading to better signal reception. Secondly, mobile antennas can be larger and, as such, more efficient. Avoid coiled antennas, as they may create wind noise while driving. The preferred location for mobile antennas is the center of the vehicle's roof, whenever possible.

When selecting antennas for multi-band radios, consider whether you primarily communicate on popular bands, such as 2m and 70cm, or less common bands, like 1.25m. Dual or tri-band antennas can provide convenience for covering multiple bands, but if specific usage is expected, opt for single-band antennas optimized for the desired frequency range.

Directional antennas are not typically recommended for portable and mobile setups, as they radiate signals in specific directions, whereas users may not always know their precise orientation. Instead, omnidirectional antennas that radiate signals uniformly in a 360° circle are generally preferred. However, directional antennas can be beneficial for stationary use in known locations with poor reception and specific communication targets.

For **fixed base stations**, directional antennas can be considered for consistent communication with parties in a particular direction. Various directional antenna patterns are available, offering different levels of directionality and signal focus. Directional antennas can substantially increase power in their main focus area, enhancing both transmission and reception.

Mitigating Interference and Noise for Clear Communication

Interference and noise are common challenges that can degrade the clarity and reliability of radio communication. For Baofeng radio users seeking to enhance range and signal quality, mitigating interference and noise becomes a crucial aspect of achieving clear communication.

Understanding the sources of interference and noise is essential for identifying and resolving communication issues. The table below highlights some common sources of interference and noise in radio communication:

Source of Interference / Noise	Description	Impact on Communication
Electromagnetic Interference (EMI)	Generated by electronic devices and power sources emitting electromagnetic fields.	Can cause signal distortion, dropouts, or complete loss of communication.

Radio Frequency Interference (RFI)	Generated by nearby radio transmitters, electronic devices, or industrial equipment operating on similar frequencies.	Can disrupt reception and result in background noise or overlapping signals.
Atmospheric Conditions (Weather-related)	Includes thunderstorms, precipitation, and temperature inversions that affect signal propagation.	Can cause signal fading, increased attenuation, and reduced communication range.
Multipath Propagation	Occurs when radio waves take multiple paths to reach the receiver, leading to signal reflections and phase shifts.	Can result in signal distortion, echo, or cancellation, affecting signal clarity.
Adjacent Channel Interference	Occurs when nearby channels interfere with the desired communication frequency.	Can lead to overlapping signals, making it challenging to isolate the intended signal.
Co-channel Interference	Occurs when two or more transmitters use the same frequency within the same geographic area.	Can cause signal interference, reducing communication range and clarity.
Electrical Noise (Man-made and Natural)	Includes man-made electrical noise from power lines and electronic devices and natural electrical noise from lightning.	Can introduce background noise and disrupt weak signals, affecting communication quality.
Intermodulation Interference	Occurs when two or more signals combine within a receiver, producing additional signals at undesirable frequencies.	Can create spurious signals, leading to distorted communication and degraded signal quality.

Strategies to Mitigate Interference and Noise

1. **Frequency Selection:** If possible, select frequencies that are less congested and free from interference. Using the appropriate repeater and avoiding crowded channels can significantly improve communication quality.

2. **Use CTCSS/DCS Tones:** Continuous Tone-Coded Squelch System (CTCSS) and Digital-Coded Squelch (DCS) tones can reduce the impact of unwanted signals. By setting CTCSS or DCS tones, the receiver will only open when receiving signals with matching codes, effectively filtering out undesired interference.

3. **Antenna Placement and Orientation:** Properly position and orient the antenna to minimize the impact of multipath propagation and maximize the signal's direct path. Raising the antenna and avoiding obstructions can improve signal reception.

4. **Signal Filters and Ferrite Beads:** Implementing signal filters and ferrite beads on power and signal cables can reduce electromagnetic interference (EMI) and radio frequency interference (RFI) caused by nearby electronic devices.

5. **Grounding and Shielding:** Ensuring proper grounding and shielding of equipment can help reduce electrical noise and interference caused by external sources.

6. **Weather Monitoring:** Pay attention to weather conditions, especially during thunderstorms, as atmospheric conditions can affect signal propagation. Avoid outdoor activities that rely on clear communication during severe weather events.

7. **Interference Hunting:** In situations where interference is persistent, consider conducting interference hunting to locate and identify the source of interference. This helps in taking appropriate measures to eliminate or mitigate the problem.

8. **Frequency Coordination:** In areas with multiple users operating on similar frequencies, coordinating frequencies and sharing information can help reduce co-channel and adjacent channel interference.

9. **Proper Equipment Maintenance:** Regularly maintain and inspect Baofeng radios and antennas to ensure optimal performance and minimize performance-related issues.

Chapter 6: Emergency Preparedness with Baofeng Radios

Developing an Effective Emergency Communication Plan

In times of crisis or emergencies, reliable and efficient communication can make a real tangible difference. Developing a well-thought-out emergency communication plan is important for individuals and groups using any sort of handheld radios. Here we outline clear procedures, designated frequencies, and communication protocols to aid communication and the exchange of information when faced with an emergency scenario.

Assessing Communication Needs and Risks

Before creating an emergency communication plan, it is crucial to assess the specific communication needs and potential risks faced by your group or community. Consider the following factors:

Factor	Description	Considerations
Location	Evaluate the geographic area where your group operates and identify potential communication challenges.	Consider topography, urban/rural environments, and natural barriers that may affect signal propagation.
Size and Structure	Determine the size and structure of your group or community to establish the scale of communication required.	Identify key roles for individuals, leaders, and communication coordinators who will be responsible for relaying information.
Potential Hazards	Identify potential emergency scenarios that your group might encounter, such as natural disasters or accidents.	Assess the severity (triage) of these hazards to prioritize communication needs.
Existing Infrastructure	Consider the availability of existing communication infrastructure, such as repeaters or communication networks.	Determine whether radios can integrate with or complement existing systems.
Power Sources	Ensure access to reliable power sources for charging radios during extended emergency situations.	Plan for backup power options like solar chargers or portable batteries.

Establishing Communication Protocols

Clear and standardized communication protocols are essential during emergencies when quick action, decision-making and precise information are critical. The following table outlines some key components of effective communication protocols:

Communication Protocol	Description	Application
Call Signs and Identifiers	Assign specific call signs or identifiers to individuals and key personnel to streamline communication.	Easily identify and address individuals during communication, enhancing clarity and organization.
Distress Signals	Establish standardized distress signals and codes that indicate an emergency or urgent situation.	Recognizable distress signals expedite emergency response and prevent misunderstandings.
Reporting Procedures	Define reporting procedures for relaying information, including the type of information to be included in reports.	Ensure consistent and comprehensive reporting of critical details during emergencies.
Check-In Procedures	Establish check-in procedures to account for all group members and verify their safety during and after an emergency.	Regular check-ins provide real-time situational awareness and aid in rescue and recovery efforts.
Priority Communications	Determine criteria for prioritizing communications based on the urgency and importance of information.	Facilitates the swift delivery of critical messages, ensuring timely responses and actions.
Chain of Command	Outline the chain of command and communication flow to ensure information flows efficiently within the group.	Clearly define roles and responsibilities to avoid communication bottlenecks and confusion during emergencies.
Backup Communication	Plan for alternative communication methods in case primary channels or devices become unavailable or compromised.	Ensure redundancy to maintain communication capabilities even in challenging situations.
Test and Practice	Regularly test and practice emergency communication	Familiarize participants with the plan, identify potential issues, and

	procedures through drills and simulations.	improve response times and coordination.

Designating Frequencies and Channels

During emergencies, communication channels may become congested. Designating specific frequencies or channels for emergency use can help prioritize critical communications and reduce interference. Ensure that all members of the group are aware of the designated emergency frequencies and channels.

Frequency/Channel Designation	Description	Application
Primary Emergency Frequency	Select a primary frequency dedicated solely to emergency communication.	This frequency will be used for crucial emergency messages and distress calls.
Secondary Emergency Frequency	Choose a backup frequency in case the primary frequency becomes unusable or congested.	This frequency acts as an alternative for emergency communication when needed.
Common Group Frequency	Designate a frequency for general group communication and coordination during emergencies.	This frequency will be used for non-emergency updates and coordination among group members.
Public Safety Frequencies	Be aware of local public safety frequencies or emergency services channels for relevant updates.	Monitor these channels for official information and announcements during emergencies.

Training and Preparedness

An effective emergency communication plan is only as strong as the individuals implementing it. Regular training sessions and drills are vital to ensure that all group members are familiar with the communication protocols and procedures. These practices enable participants to handle emergency situations confidently and efficiently.

Training and Preparedness	Description	Application

Communication Drills	Conduct regular communication drills and simulations to practice emergency procedures.	Drills improve response times, foster teamwork, and identify areas that need improvement.
Equipment Familiarization	Ensure all members are familiar with handheld radios and other communication devices.	Familiarity with equipment reduces confusion during emergencies and promotes effective usage.
Role-Specific Training	Train individuals in specific roles, such as communication coordinators and call sign operators.	Role-specific training enhances efficiency and coordination among designated personnel.
Emergency Plan Reviews	Periodically review and update the emergency communication plan based on lessons learned.	Regular reviews help adapt the plan to changing circumstances and address any identified shortcomings.

Connecting with Emergency Services and Networks

During emergencies, effective communication with the relevant emergency services can be pivotal for obtaining assistance, coordinating resources, and staying informed about the situation. Radio operators have access to various emergency frequencies and networks, making them valuable tools for connecting with the appropriate authorities and receiving vital updates.

The proficiency in operating a ham radio transcends being a mere contingency skill for survivalists, preppers, or first responders. In scenarios where conventional communication infrastructures may are down, being familiar with radio operations could prove invaluable in an emergency. Possessing a practical understanding of how Baofeng radios operate, along with knowledge of their emergency frequencies, can make the difference in a life-threatening situation.

Emergency Frequencies in the US

In the United States, several designated frequencies are reserved for emergency communications. These frequencies are monitored by emergency services, law enforcement agencies, and first responders. Radio operators can program these frequencies into their devices to access emergency services and receive critical updates during emergencies. Below are some important emergency frequencies in the US:

Frequency	Designation	Purpose
34.90 MHz	National Guard Emergency Channel	Nationwide emergency communication for the National Guard.
39.46 MHz	Police Emergency Channel	Emergency communication channel for local and state police.
47.42 MHz	Red Cross Relief Operations Channel	Nationwide channel for Red Cross relief operations.
52.525 MHz	Six-Meter Band FM Calling Frequency	Used by ham radio operators, especially during exceptional conditions.
121.50 MHz	International Aeronautical Emergency Frequency	Emergency communication for aircraft.
138.225 MHz	FEMA Disaster Relief Channel	Primary FEMA channel for disaster relief operations.
146.52 MHz	Two-Meter Band Simplex Channel	Ham radio frequency for non-repeater communications.
151.625 MHz	Mobile Business and Events Channel	Used by mobile businesses, exhibitions, trade shows, sports teams.
154.28 MHz	Local Fire Department Emergency Channel	Emergency communication channel for fire departments.
155.160 MHz	Local and State Agency Search and Rescue	Used by agencies for search and rescue operations.
155.475 MHz	Local and State Police Emergency Channel	Emergency communication channel for police.
156.75 MHz	International Maritime Weather Alerts	Maritime weather alerts and updates.
156.80 MHz	International Maritime Distress and Safety	Distress calls, initial communication, and safety

		on the water.
162.40-162.55 MHz	NOAA Weather Broadcasts	Range of channels for NOAA weather broadcasts and bulletins.
163.4875 MHz	National Guard Emergency Channel	Nationwide emergency communication for the National Guard.
163.5125 MHz	Armed Forces Disaster Preparedness Channel	National disaster preparedness frequency for the armed forces.
164.50 MHz	Department of Housing and Urban Development	National communication channel for HUD.
168.55 MHz	Federal Civilian Agency Emergency Channel	Federal civilian agency channel for emergencies.
243.00 MHz	Military Aviation Emergency Channel	Emergency channel for military aviation.
259.70 MHz	Space Shuttle Re-entry and Landing (UHF)	Frequencies for Space Shuttle re-entry and landing.
296.80 MHz	Space Shuttle Re-entry and Landing (UHF)	Frequencies for Space Shuttle re-entry and landing.
311.00 MHz	U.S. Air Force In-Flight Channel	Active channel for U.S. Air Force in-flight communication.
319.40 MHz	U.S. Air Force In-Flight Channel	Active channel for U.S. Air Force in-flight communication.
317.70 MHz	U.S. Coast Guard Aviation Channel	Active channel for U.S. Coast Guard aviation.
317.80 MHz	U.S. Coast Guard Aviation Channel	Active channel for U.S. Coast Guard aviation.
340.20 MHz	U.S. Navy Aviation Channel	Active channel for U.S. Navy aviation.

409.20 MHz	Interstate Commerce Commission Communication	National communication channel for the Interstate Commerce Commission.
409.625 MHz	Department of State Communication Channel	National communication channel for the Department of State.
462.675 MHz	GMRS Emergency Communication Channel	General Mobile Radio Service channel for emergency communication.

It is important to note that using emergency frequencies requires a licensed operator or authorization during emergencies. In situations where you are not a licensed amateur radio operator, the use of a radio device on emergency frequencies should be done responsibly and with caution.

International Emergency Frequencies and Networks

For travelers and individuals outside the US, understanding international emergency frequencies and networks is essential. Different countries may allocate specific frequencies for emergency communication. Below are some international emergency frequencies:

Maritime Mobile Service Frequencies

Frequency (kHz)	Designation	Purpose
2182	Medium Range Maritime Voice	Used for distress calls and communication in medium-range maritime areas.
4125	HF Maritime Voice	HF frequency for long-distance distress calls in maritime operations.
6215	HF Maritime Voice	HF frequency for long-distance distress calls in maritime operations.
8291	HF Maritime Voice	HF frequency for long-distance distress calls in maritime

		operations.
12290	HF Maritime Voice	HF frequency for long-distance distress calls in maritime operations.
16420	HF Maritime Voice	HF frequency for long-distance distress calls in maritime operations.
156.8 MHz	Marine VHF Radio Channel 16	Used for short-range maritime distress and emergency communication.
406-406.1 MHz	Cospas-Sarsat Satellite SAR System	International satellite-based search and rescue distress alert system.

Digital Selective Calling (DSC) Frequencies

Frequency (MHz)	Designation	Purpose
2.1875	DSC Frequency	Used for digital selective calling in maritime communications.
4.2075	DSC Frequency	Used for digital selective calling in maritime communications.
6.312	DSC Frequency	Used for digital selective calling in maritime communications.
8.4145	DSC Frequency	Used for digital selective calling in maritime communications.
12.577	DSC Frequency	Used for digital selective calling in maritime communications.
16.8045	DSC Frequency	Used for digital selective calling in maritime communications.
156.525 MHz	Marine VHF Radio Channel 70	Used for digital selective calling in marine VHF communications.

Aeronautical and Search And Rescue (SAR) Frequencies

Frequency (MHz)	Designation	Purpose
121.5	Civilian Aircraft Emergency	Civilian aircraft emergency frequency and international air distress.
243	NATO Military Aircraft Emergency	NATO military aircraft emergency frequency.
123.1	Aeronautical Auxiliary Frequency	International voice for coordinated SAR operations.
138.78	U.S. Military Voice SAR	U.S. military voice SAR on-the-scene use and direction finding (DF).
155.16	SAR Frequency	Search and rescue communication frequency.
172.5	U.S. Navy Emergency Sonobuoy	U.S. Navy emergency sonobuoy communications and homing use.
282.8	Joint/Combined On-The-Scene	On-the-scene voice and DF frequency used throughout NATO.
406-406.1 MHz	Cospas-Sarsat Satellite SAR System	International satellite-based search and rescue distress alert system.

In addition to specific frequencies, there are international networks and organizations dedicated to emergency communications. For instance, the International Radio Emergency Support Coalition (IRESC) is a global organization of amateur radio operators who provide emergency communication assistance during disasters and crises. The Amateur Radio Emergency Service (ARES) is another network in the US that coordinates amateur radio operators to assist with emergency communications at the local level.

Connecting with Emergency Services and Networks

Connecting with emergency services and networks using Ham radios requires proper programming and understanding of the emergency frequencies and protocols. Here are some key steps to connect with emergency services:

1. **Programming Emergency Frequencies:** Ensure that you have the necessary frequencies programmed into your Ham radio. Consult local authorities or amateur radio organizations for information about emergency frequencies in your region.

2. **Monitoring NOAA Weather Radio:** Keep your radio tuned to the NOAA Weather Radio frequency (162.400 MHz) to receive weather alerts, watches, and warnings issued by the National Weather Service.

3. **Listening for Emergency Broadcasts:** Monitor the designated emergency frequencies for updates and announcements from emergency services and organizations.

4. **Contacting Authorities:** In an emergency situation, use the appropriate emergency frequencies to contact local authorities or emergency services for assistance.

5. **Participating in Emergency Networks:** If you are a licensed amateur radio operator, consider participating in amateur radio emergency networks such as ARES or IRESC to provide communication support during emergencies.

6. **Coordinating with Local Clubs:** Reach out to local amateur radio clubs or organizations to learn about emergency communication practices and participate in training exercises and drills.

7. **Maintaining Responsible Usage:** Use emergency frequencies responsibly and avoid causing unnecessary interference. Respect the protocols and rules associated with these frequencies.

Above all, it is important to remember that emergency frequencies are critical resources used by first responders and emergency services during crises. Ham radio users should use these frequencies responsibly and with the intention of contributing to emergency response efforts when authorized or licensed to do so.

Maintaining and Testing Your Radio for Readiness

It is important to regularly maintain and test your Baofeng radio to ensure reliability during emergency situations. Your radio should be working at optimal performance and not something you want causing problems in an emergency. Regularly maintenance should not be something you postpone or neglect. Not carrying out regular maintenance can lead to problems and degrade capacity to communicate, which might jeopardize your ability to stay connected when you need it the most.

1. **Battery Maintenance:**

   The battery is the lifeblood of your radio. Regularly inspect the battery for signs of wear, damage, or corrosion. Ensure that the battery contacts are clean and free from debris, as a poor connection can lead to power issues. It is advisable to have spare fully charged batteries available, especially during extended trips or emergencies, to ensure uninterrupted communication.

2. **Antenna Check:**

The antenna is a critical component that affects the radio's range and signal quality. Examine the antenna regularly for any signs of damage, such as bends or cracks. Ensure that it is securely attached to the radio. If you use a removable antenna, consider upgrading to a higher-quality antenna to enhance performance. Chapter 5 provides an overview of common Baofeng radio antenna types and their characteristics.

3. **Firmware and Software Updates:**

Periodically check for firmware and software updates for your Baofeng radio model. Manufacturers may release updates to improve functionality, address bugs, and enhance performance. Follow the manufacturer's guidelines to safely update your radio's firmware and programming software, if applicable.

4. **Functionality Test:**

Regularly test the functionality of your Baofeng radio to ensure all features and buttons are working correctly. Verify that the display is clear and legible, and the keypad responds as expected. Conduct a test transmission to confirm that the microphone and speaker are functioning properly.

5. **Programming Validation:**

If you have saved specific frequencies and channels into your Baofeng radio, periodically validate these to ensure they are accurate and up to date. Changes in regulations or emergency frequencies may happen at any time.

6. **Range Testing:**

Conduct range testing in different environments to understand the actual communication range of your Baofeng radio. Factors such as terrain, obstructions, and atmospheric conditions can affect the radio's range. Knowing the effective range will help you make informed decisions during outdoor trips or during emergencies.

By regularly maintaining and testing your Baofeng radio, you can have confidence in its readiness to perform when needed. Regular checks and being proactive will help avoid unexpected issues and ensure that your communication device remains a reliable tool for emergency preparedness.

Chapter 7: Real-Life Scenarios and Case Studies

Case Studies of Successful Baofeng Radio Use in Survival Situations

In this first section, we present some real-life case studies that demonstrate the effectiveness of having handheld radios in survival situations. These case studies illustrate how radios have been instrumental in connecting individuals with emergency services and facilitating successful rescue operations during critical times. I have purposefully removed any identifiable information to ensure anonymity.

Case Study 1: Hiking Emergency in the Wilderness

Location: Appalachian Trail, United States

Time: May 15, 2002

Scenario: A group of hikers was traversing a challenging section of the Appalachian Trail when one of the members suffered a severe ankle injury, rendering them immobile. As the incident happened in a remote area with no cellphone signal, the group was isolated and unable to call for help.

Response: Fortunately, one of the hikers had a Baofeng UV-5R handheld radio equipped with programmed emergency frequencies. Recognizing the urgency of the situation, the group activated the Baofeng radio's distress signal on the designated emergency frequency, 155.160 MHz, used for search and rescue operations. A local amateur radio operator who monitored the frequency as part of an emergency response network picked up the signal.

Emergency Services Involved: The amateur radio operator quickly relayed the distress call to the local Search and Rescue (SAR) team and provided them with the GPS coordinates obtained from the hikers' radio transmission.

Outcome: Thanks to the fact one of the members carried a radio and their quick response time to send out a distress signal, efficient communication was made between the hikers and the amateur radio operator. The SAR team was then able to pinpoint the injured hiker's location and mobilize a rescue mission promptly. The injured hiker received medical attention within hours of the incident, and the rest of the group was safely guided out of the wilderness.

Case Study 2: Severe Weather Emergency

Location: Coastal Area, Japan

Time: September 10, 2007

Scenario: A typhoon struck a rural area of the coast of Japan, causing overall damage, flooding, and power outages. Communication infrastructure was severely affected, leaving many residents stranded and isolated.

Response: A local community prepared for such emergencies and equipped their disaster preparedness kit with UV-82HP radios, preprogramed with NOAA weather radio frequencies and the emergency channel 168.55 MHz. During the typhoon, the community's designated emergency response team used the radios to maintain communication among key personnel across various vilages, coordinate relief efforts, and monitor weather updates.

Emergency Services Involved: Due to the extensive damage, local emergency services were overwhelmed and had limited resources to respond immediately. The community's reliance on their Baofeng radios allowed them to communicate and assess the situation on the ground efficiently.

Outcome: The community's proactive use of the handheld radios enabled them to take charge of thei emergency situation and be on top of it. They were able to provide critical information to local authorities and facilitated the quick evacuation and assistance of the affected residents across multiple vilages. As they had pre-saved the NOAA weather radio frequencies on their radio devices, they were able to stay informed about the typhoon's movements and potential risks, helping them make well-informed decisions to safeguard lives and property.

Case Study 3: Off-Road Adventure Emergency

Location: Outback, Australia

Time: November 20, 2015

Scenario: A group of off-road enthusiasts was exploring a remote region in the Australian Outback when one of their vehicles broke down. They were far from any major roads or towns, and the scorching desert heat made the situation more perilous.

Response: Recognizing the importance of reliable communication in remote areas, the group had equipped their vehicles with BF-F8HP radios programmed with both local repeater frequencies and the emergency channel 27.555 MHz (CB radio emergency channel).

Emergency Services Involved: The group activated their radios and attempted to reach out to any other travelers or nearby stations on the CB radio emergency channel. After several attempts, they made contact with another off-road group who were within range.

Outcome: The group was able to relay their location and situation to the nearby off-road group, who, in turn, contacted the local authorities and provided them with the necessary details for a rescue operation. The group that had been become stranded received assistance from the local rescue team within hours and were retrieved from the remote region.

These case studies highlight the critical role radios play in emergency situations. These devices connect individuals with emergency services and can greatly expedite rescue operations. Properly equipped and used with foresight, radios can be indispensable tools for enhancing safety and communication during critical times in various survival scenarios.

Practical Exercises to Improve Communication Skills

Knowing how to use your radio efficiently and being familiar with common communication protocols can be of great help to effectively conveying information during emergencies and outdoor adventures. These exercises will help users build confidence, practice emergency procedures, and improve their radio communication capabilities.

Exercise 1: Radio Familiarizatio

Objective: Familiarize yourself and your team with the each of the radio's features, buttons, and functionalities.

Procedure:

1. Gather your team and distribute radios to each member.

2. Review the user manual together to understand the radio's various functions and settings.

3. Practice turning the radio on and off, adjusting the volume, and navigating through the menu options.

4. Demonstrate how to program frequencies and channels using CHIRP software or manual programming methods.

5. Conduct sample transmissions among team members to ensure everyone is comfortable with basic communication procedures.

Exercise 2: Scenario-based Simulations

Objective: Simulate emergency scenarios to practice effective radio communication under different conditions.

Procedure:

1. Create a hypothetical emergency scenario such as hikers getting lost in the wilderness, medical emergencies, or vehicle breakdowns.

2. Assign roles to team members, including a designated leader, communicator, and responder.

3. Employ radios to coordinate responses and relay critical information between team members.

4. Emphasize clear and concise communication, using proper radio etiquette, and avoiding jargon or unnecessary chatter.

5. Debrief and review the exercise afterward to identify areas for improvement. Discuss among you different ways you could enhance communication.

Exercise 3: Range Testing and Signal Clarity

Objective: Determine the effective range of your radios and optimize signal clarity.

Procedure:

1. Choose a safe location with a clear line of sight and minimal obstructions.

2. Select two team members to act as communicators, with one stationed at the starting point and the other at a predetermined distance away.

3. Have the communicators establish contact and conduct transmissions, periodically adjusting the radio's settings (e.g., transmit power, squelch level) to optimize signal clarity.

4. Measure the distance at which communication remains reliable and record the findings for future reference.

5. To understand how various conditions impact radio range and signal quality, repeat the exercise in different environments, (e.g. Urban areas, forests, or hilly terrains).

Exercise 4: Cross-Band Repeater Operation

Objective: Explore the functionality of cross-band repeaters to extend radio communication range.

Procedure:

1. Familiarize yourself with the cross-band repeater feature on a compatible radio models.

2. Set up a cross-band repeater using two radios, with one configured as the transmitter and the other as the receiver.

3. Position the receiver at a higher elevation or a location with better line-of-sight to improve radio coverage.

4. Test communication from various distances and observe the extended range achieved through the repeater operation.

5. Discuss the pros and cons of cross-band repeaters in relation to their use in extending communication capabilities during emergencies.

Exercise 6: Prowords and Phonetic Alphabet Proficiency

Objective: Develop proficiency in using prowords and the NATO phonetic alphabet to enhance clarity and precision during radio communication. Please check Chapter 2 for more details.

This exercise will focus on:

- Using Prowords: Begin with familiarizing yourself with the concept of prowords.

- Familiarizing with NATO Phonetic Alphabet: Introduce the NATO phonetic alphabet.

Procedure:

- Team Composition: Divide the hikers into two groups - Group A and Group B. Group A will represent the main group, and Group B will be the separated hikers.
- Communication Equipment: Equip each hiker in both groups with a handheld radio. Ensure that all radios are pre-programmed with the same frequencies and channels for communication.
- Proword and Phonetic Alphabet Introduction: Before starting the exercise, brief all hikers about the importance of using prowords and the NATO phonetic alphabet for clear and concise communication. Provide a handout with a list of prowords and the phonetic alphabet for easy reference.
- Separation and Communication: Have Group B members move a short distance away from Group A, simulating the separation scenario. Instruct Group B to use the radios to initiate contact with Group A and relay their current location.
- Using Prowords and Phonetic Alphabet: Group B members should use prowords like "Break, Break, Break" to indicate an urgent message and "This is" to introduce themselves. They should then use the NATO phonetic alphabet to spell out their exact location and any landmarks nearby.
- Responding with Prowords: Group A, upon receiving the message from Group B, should acknowledge the message using prowords like "Roger" to indicate they received the message and "Over" to signal the end of their transmission.

- Coordinating Reunion: Once Group A has received Group B's location, they should provide clear instructions on how to reunite. For example, they might say, "Head South for 100 meters, then look for the large oak tree on your left. We will meet you there. Over."
- Real-Time Feedback: Observe and provide real-time feedback on each group's use of prowords and phonetic alphabet. Encourage both groups to practice concise and accurate communication while incorporating the introduced terminology.
- Repeat and Switch Roles: Have the groups switch roles, allowing Group B to become the main group and Group A to be the separated hikers. Repeat the exercise to reinforce the importance of proficient communication in both scenarios.
- Debrief and Discussion: After completing the exercise, gather all hikers for a debriefing session. Discuss challenges faced, successful communication instances, and areas for improvement. Encourage open discussions on how using prowords and the phonetic alphabet enhanced communication efficiency.

Understanding Essential Radio Frequencies for Emergencies

Although we have covered many of the emergency networks on Chapter 6 "Emergency Preparedness with Baofeng Radios", in this section, we will take a step back and talk about general radio frequencies used in emergencies. Having knowledge of the specific frequencies used by emergency services and agencies is definitely important for a providing quick and effective response during adverse situations. Here, we explore key frequency bands and their associated emergency services, allowing users to make informed decisions when configuring their handheld radios for emergency preparedness.

VHF and UHF Bands

The Very High Frequency (VHF) and Ultra High Frequency (UHF) bands are commonly utilized for emergency communication due to their line-of-sight propagation characteristics, making them suitable for local and regional communication.

VHF Band (30-300 MHz): This band includes frequencies often used by local emergency services such as police, fire departments, and emergency medical services (EMS). One notable frequency is 155.160 MHz, used for search and rescue operations. Additionally, ham radio operators utilize the 2-meter band, which operates between 144-148 MHz, and the 70cm band, which operates between 420-450 MHz, for emergency communication.

UHF Band (300-3000 MHz): Within the UHF band, frequencies around 460-470 MHz are commonly used by public safety agencies for communication. Some regions allocate specific UHF frequencies for amateur radio operators to support emergency response efforts.

HF Band

The High Frequency (HF) band is valuable for long-distance communication, especially during disasters when standard communication infrastructure may be compromised. HF radio waves can bounce off the ionosphere, allowing communication over long distances without direct line-of-sight.

HF Emergency Frequencies: The Citizens Band (CB) radio frequencies around 27 MHz and the 40-meter band (7.0-7.3 MHz) are often designated for emergency use by ham radio operators. The 20-meter band (14.0-14.35 MHz) and 80-meter band (3.5-4.0 MHz) are also popular choices for emergency communication.

NOAA Weather Radio Frequencies

NOAA Weather Radio broadcasts critical weather updates and alerts, including severe weather warnings, watches, and advisories.

162.40-162.55 MHz: This band is reserved for NOAA Weather Radio broadcasts. Baofeng radios with NOAA weather alert features can receive these broadcasts and provide users with real-time weather information to aid in making informed decisions during outdoor activities and emergencies.

Satellite-Based Emergency Frequencies

Cospas-Sarsat is an international satellite-based search and rescue system that detects and locates distress beacons activated during emergencies.

406 MHz: Distress beacons and personal locator beacons (PLBs) transmit distress signals on this frequency, enabling search and rescue authorities to locate those in distress quickly.

Military and Government Frequencies

Certain frequencies are reserved for military and government use during emergencies and disasters.

Military Aviation Frequencies: Frequencies like 121.5 MHz and 243 MHz are designated for civilian aircraft emergencies and military aviation emergencies, respectively. While 121.5 MHz is no longer actively monitored by Cospas-Sarsat, it is still in use for general aviation emergency communications.

Government Agency Frequencies

Frequencies in the 400 MHz range are often used by government agencies for emergency operations and coordination.

Being aware of the specific bands is certainly useful and can make a difference in an emergency situation. However, it is crucial to note that some frequencies may require specific licenses or certifications to transmit legally. Therefore, users should familiarize themselves with local regulations and adhere to responsible radio practices to ensure they are not impeding emergency services.

Chapter 8: Troubleshooting & Maintenance

Identifying and Resolving Common Baofeng Radio Issues

Baofeng radios are know for their versatility and affordability, but they can encounter various issues during their lifetime of use. In this section, we address common problems affecting Baofeng radios:

1. **Fixing a Baofeng UV-5R that Stops Receiving after CHIRP File is Uploaded:**

   Issue: After uploading a CHIRP .img file into a new Baofeng radio, users may find that the radio stops receiving signals.

   Solution: Check the squelch settings on the Baofeng radio. It's possible that the squelch is set too high, causing the radio to mute incoming signals. Adjust the squelch to an appropriate level to regain reception.

   Ensure that the radio's frequency and channel settings are correctly configured in the CHIRP software and match the desired frequencies for your location. Incorrect settings can lead to communication issues.

2. **Common Error Messages:**

   Baofeng radios may display various error messages, indicating potential problems during operation.

   a. **"Err" or "Error":** This message typically indicates that the entered frequency is out of the radio's supported range. Verify that the frequency is within the supported range for the specific Baofeng model.

   b. **"CH Error":** This error may appear if the selected channel does not have a valid frequency programmed. Double-check the channel settings and ensure they correspond to the intended frequencies.

   c. **"Busy Channel":** This message suggests that the selected channel is currently in use by another radio. Choose an alternate channel for communication.

3. **Erratic Behavior and Recovery (Baofeng/Pofung UV5R and F8HP Series Radios):**

   Issue: Baofeng radios may exhibit erratic behavior, such as spontaneous resets or malfunctioning controls.

Solution: Perform a factory reset: This may help resolve software-related issues causing erratic behavior. Refer to the user manual for instructions on how to perform a factory reset for your specific model.

Check for firmware updates: Some issues may be addressed through firmware updates released by Baofeng. Check the manufacturer's website for the latest firmware version and follow the update instructions.

4. **What to Do if CHIRP Does Not List Your Radio.**

   Issue: Some users may find that their specific Baofeng radio model is not listed in CHIRP software.

   Solution: Check for CHIRP updates: Ensure that you are using the latest version of CHIRP, as new updates often include support for additional radio models.

   Manually program your radio: If CHIRP does not support your radio model, resort to manual programming using the keypad. Refer to the user manual for step-by-step instructions on manual programming.

5. **What to Do When the Battery Pack Does Not Charge:**

   Issue: Users may encounter issues with their Baofeng radio battery not charging properly or holding a charge.

   Solution: Inspect the battery contacts: Ensure that the battery contacts on both the radio and the battery pack are clean and free from debris. Use a soft cloth or cotton swab to clean if necessary.

   Verify the charging cable and adapter: Confirm that the charging cable and adapter are functional by testing them with other devices or using a different cable/adapter to charge the radio.

   Replace the battery: If the battery fails to hold a charge or is damaged, consider replacing it with a genuine Baofeng replacement battery.

6. **What to Do if the Radio Does Not Power On:**

   Issue: The Baofeng radio may not power on despite having a charged battery.

   Solution: Check the battery connection: Ensure that the battery is properly inserted into the radio and that the contacts are making a secure connection.

Try a different battery: If possible, test the radio with another fully charged battery to rule out battery-related issues.

Perform a hard reset: Some issues may be resolved by performing a hard reset, which involves removing the battery and holding the power button down for several seconds before reinserting the battery.

7. **What to Do if the Programming Cable Does Not Detect the Radio:**

Issue: The programming cable may fail to detect the Baofeng radio when connected to a computer.

Solution: Check the cable connection: Ensure that the programming cable is securely plugged into both the Baofeng radio and the computer's USB port.

Install the correct driver: Verify that the appropriate programming cable driver is installed on the computer. Check the manufacturer's website for driver downloads and installation instructions.

Try a different USB port: Test the programming cable on different USB ports to rule out any port-related issues.

8. **What to Do if the Programming Cable Does Not Work:**

Issue: The programming cable may fail to function properly, preventing communication between the radio and the computer.

Solution: Check cable integrity: Inspect the programming cable for any visible damage or wear. If the cable is damaged, consider replacing it with a new one.

Use genuine programming cable: Ensure that you are using a genuine Baofeng programming cable specifically designed for your radio model.

Verify cable compatibility: Confirm that the programming cable is compatible with your computer's operating system and version.

9. **What to Do if the Radio Programming Software Won't Detect the Cable or COM Port:**

Issue: The programming software may fail to recognize the programming cable or the COM port used for communication.

Solution: Restart the software and computer: Sometimes, a simple restart of both the programming software and the computer can resolve connection issues.

Reinstall software and drivers: Uninstall and then reinstall the programming software and the corresponding programming cable drivers to refresh the connection.

Try a different USB port and cable: Test the programming cable on different USB ports and use an alternate cable if available.

Check device manager (Windows): In Windows, open the Device Manager to see if the programming cable is recognized under "Ports (COM & LPT)." If it appears with a warning sign, troubleshoot the driver or try a different USB port.

Proper Maintenance to Prolong Radio Lifespan

Proper maintenance of Baofeng radios is crucial to ensure their longevity and reliable performance. In this section, we will explore essential maintenance practices that users should incorporate into their routine to prolong the lifespan of their radios.

Regular Cleaning

External Surfaces: Baofeng radios are often used in outdoor environments, where they can accumulate dust, dirt, and grime. Regularly clean the external surfaces of the radio using a soft, lint-free cloth dampened with water or a mild, non-abrasive cleaning solution. Avoid using harsh chemicals that may damage the radio's casing.

Antenna and Connectors: Clean the antenna and connectors periodically to prevent signal interference and ensure proper radio operation. Use a soft brush or cotton swab to remove any debris or oxidation on the antenna and connectors.

Battery Maintenance

Charging Practices: Follow the manufacturer's guidelines for charging the battery. Overcharging or using incompatible chargers can damage the battery and reduce its lifespan. Avoid using the radio while it is charging to prevent excessive heat buildup.

Battery Storage: If the radio will not be used for an extended period, store the battery in a cool, dry place at approximately 50% charge. This helps prevent over-discharge and maintains battery health during storage.

Battery Calibration: Periodically calibrate the battery to ensure accurate power level readings. Fully discharge the battery and then charge it to 100% to recalibrate the battery's internal capacity measurements.

Proper Handling and Storage

Protective Cases: Use a durable and well-fitted protective case to safeguard the radio from impacts, scratches, and exposure to the elements during outdoor activities. The case should allow easy access to essential functions and buttons.

Temperature Considerations: Avoid exposing the radio to extreme temperatures, especially high heat or freezing cold, as it can negatively impact battery life and overall radio performance.

Storage Conditions: When not in use, store the radio in a dry, cool environment away from direct sunlight. Avoid storing the radio in humid or damp locations to prevent damage to internal components.

Antenna Care

Securely Attach the Antenna: Ensure that the antenna is properly attached to the radio and tightened securely to prevent signal loss and enhance transmission and reception performance.

Inspect for Damage: Regularly inspect the antenna for any signs of wear, bending, or damage. A damaged antenna can negatively impact signal strength and communication range.

Firmware and Software Updates

Stay Updated: Check for firmware and software updates provided by the manufacturer. These updates often include bug fixes, feature enhancements, and improvements to the radio's performance and stability.

Follow Update Instructions: Before updating the firmware or programming software, carefully read the manufacturer's instructions to avoid potential issues during the update process.

Perform Regular Testing

Functionality Checks: Periodically conduct functional tests to ensure that all features and buttons on the radio are working correctly. This includes checking the keypad, LCD display, emergency features, and audio quality.

Signal Testing: Test the radio's transmission and reception capabilities by communicating with another radio user at varying distances. This helps assess the radio's performance and range.

Seek Professional Assistance

Authorized Service Centers: If the radio experiences technical issues beyond user troubleshooting abilities, seek assistance from an authorized Baofeng service center or a qualified radio technician. Attempting to repair complex issues without expertise may cause further damage.

By incorporating these proper maintenance practices into their routine, users can prolong the lifespan and ensure reliable performance of their Baofeng radio.

Recommended Accessories and Upgrades

To enhance the functionality and performance of Baofeng radios, users can consider investing in various accessories and upgrades. These additions not only improve communication capabilities but also contribute to the overall reliability and convenience of the radio. In this section, we will explore some recommended accessories and upgrades that Baofeng radio users may find beneficial.

High-Quality Antennas

One of the most impactful upgrades for Baofeng radios is a high-quality aftermarket antenna. The stock antennas provided with Baofeng radios are often sufficient for basic use, but upgrading to a more efficient and specialized antenna can significantly improve signal range and clarity.

- **Stubby Antennas:** Compact stubby antennas like the Diamond SRH805S minimize bulk and interference, making them ideal for local, short-range communications.

- **Longer Antennas:** Longer antennas like the Nagoya NA-771 offer increased signal range, allowing users to reach distant repeaters and stations. These antennas are well-regarded for their performance and are often preferred for emergency situations that demand extended transmission reach.

- **Flexible Antennas:** The "Signal Stick antenna", which is flexible and can be easily looped, proves beneficial for storing in bags or plate carriers without impeding mobility.

- **Roll-up Antennas:** The N9TAX roll-up antenna provide excellent transmission distance without sacrificing portability. These antennas are lightweight and can be deployed when stationary, making them well-suited for go-bags and emergency scenarios.

Extended Battery Packs

In certain situations where communication needs may extend over an extended period, consider using an extended battery like the Baofeng Extended 3800 mAh battery, which provides significantly longer battery life. This battery also features a DC barrel plug input, allowing convenient charging without requiring the bulky dock. With an adapter cable, users can charge their Baofeng radio from various power sources with USB ports, such as solar panels or battery packs, ensuring continuous power supply during emergencies.

Remote Speaker Microphones (RSM)

Remote speaker microphones, also known as shoulder microphones or hand microphones, offer added convenience during communication. These accessories allow users to clip the microphone to their clothing and communicate without holding the radio directly. RSMs come with built-in speakers, which improve audio clarity and volume, making it easier to communicate in noisy or windy environments. Consider something like the QHM22 Speaker Mic.

USB Charging Cable

The conventional charging method involves using the port, which is limited to connection with a wall outlet. However, the acquisition of a USB charging cable expands this limitation, allowing for in-car charging via an inverter, or even enabling solar panel charging.

Belt Clips and Pouches

To ensure easy access to the radio while keeping hands free during outdoor activities, belt clips and holsters are indispensable. The ITS Tactical 10-4 Radio Pouch offers a versatile solution. The pouch can be mounted on belts, PALS/MOLLE webbing, or shoulder straps, providing easy access to all radio controls. It securely holds a BaoFeng radio with an extended battery, and an optional lanyard and retraction system can prevent the radio from getting lost during movements.

Carrying Cases and Protective Covers

Investing in high-quality carrying cases and protective covers is essential for protecting Baofeng radios during rugged outdoor adventures. These accessories shield the radio from impacts, scratches, and exposure to dust and water, prolonging the radio's lifespan and preserving its optimal performance.

External Power Amplifiers

For users seeking extended communication range, especially in challenging terrain, external power amplifiers can significantly boost the radio's output power. These devices amplify the radio signal, resulting in improved transmission range and signal penetration.

External Antennas for Vehicle Use

When using Baofeng radios in vehicles, an external antenna is essential to overcome the Faraday cage effect that blocks radio signals within the car. The Nagoya UT-72 magnetic mount antenna is highly effective and attaches firmly to the vehicle's roof. To ease cable management, users may consider using BNC adapters to switch antennas effortlessly. A car mount, typically designed for smartphones, can be used to securely hold the radio in place for easy access while driving.

Battery Eliminator for Vehicle Charging

For extended vehicle trips, users can opt for the Baofeng battery eliminator, which connects to the radio's dummy battery and draws power directly from the vehicle's cigarette lighter socket. This accessory spares the battery from wear and tear and ensures a continuous power supply while on the road.

Chapter 9: Legal and Regulatory Compliance

Essential Knowledge on FCC Regulations for Baofeng Radio Use

Operating Baofeng radios, like any other two-way radio, requires compliance with the regulations set forth by the Federal Communications Commission (FCC). The FCC is the regulatory authority in the United States responsible for managing and overseeing the use of radio frequencies.

It is essential for radio operators, including Baofeng enthusiasts, to have a good understanding of these regulations to ensure legal and responsible radio communication.

Part 95 Regulations:

Baofeng radios are classified as Part 95 devices under the FCC regulations. Part 95 encompasses various radio services, including Family Radio Service (FRS), General Mobile Radio Service (GMRS), and Multi-Use Radio Service (MURS). It is essential to understand which specific service your Baofeng radio operates under, as the rules and requirements differ for each.

Part 97: Amateur Radio Service:

For those who have obtained an Amateur Radio License, also known as a ham license, they can use Baofeng radios on frequencies allocated to the Amateur Radio Service. This service allows licensed operators to communicate with other Ham radio enthusiasts locally, nationally, and even globally using a wide range of frequencies across different bands.

Amateur radio operations are governed by Part 97 of the FCC rules and regulations. Complying with these rules ensures that radio communications remain efficient, respectful, and interference-free. Some key provisions of Part 97 include:

1. Frequency Privileges: Each amateur radio license class provides specific frequency privileges. Operators must strictly adhere to these privileges and refrain from using frequencies outside their authorized bands.

2. Emission Types: The FCC regulates the types of emissions (modulation techniques) permitted in different frequency bands. For example, voice communication is typically permitted using amplitude modulation (AM) and frequency modulation (FM), while Morse code transmissions may use continuous wave (CW) emissions.

3. Station Identification: Amateur radio operators are required to identify their stations at regular intervals during transmissions. As mentioned in previous chapters, this

identification is typically done with their unique call signs which help identify the source of transmissions.

4. Prohibited Communications: FCC regulations prohibit certain types of communications on amateur radio frequencies, such as commercial activities, encrypted messages, and broadcasting music or entertainment content.

5. Operating Etiquette: Good operating practices are crucial for maintaining efficient and courteous communication on the airwaves. Ham radio operators should adhere to standard operating procedures (SOPs) and exhibit proper radio etiquette. This includes using clear and concise language, avoiding unnecessary transmissions, and being considerate of other operators.

6. Interference Mitigation: Ham radio operators should ensure that their transmissions do not cause harmful interference to other licensed radio users. If interference is detected, operators must take corrective actions to eliminate or minimize it.

Unlicensed Use:

Certain frequencies in Baofeng radios, such as those allocated for CBRS, FRS and MURS, do not require an individual license for use. These are known as unlicensed frequencies, and anyone can use them as long as they comply with the power limits, channel restrictions, and other rules specified by the FCC. However, it is crucial to note that use on certain frequencies, such as GMRS, requires a license.

Transmit Power Limits:

The FCC specifies the maximum power output (transmit power) allowed for each radio service. For instance, on FRS frequencies, the maximum transmit power is limited to 0.5 watts, while GMRS radios can operate at higher power levels with a valid license. Adhering to these power limits ensures that radio signals do not cause interference with other communication systems.

Type-Acceptance and Certification:

All Baofeng radios must undergo certification and type-acceptance processes to comply with FCC regulations. This ensures that the radios meet specific technical standards and do not cause harmful interference to other radio services. To ensure you meet regulatory requirements it is important to always purchase FCC-certified Baofeng radios from authorized sellers.

Prohibited Frequencies and Activities:

The FCC strictly prohibits the use of Baofeng radios for unauthorized purposes or on unauthorized frequencies. For example, using Baofeng radios for illegal activities, such as jamming or unauthorized broadcasting, is strictly prohibited and subject to severe penalties. Moreover, using the radio on frequencies not allocated for the specific radio service can lead to interference and may result in legal consequences.

Antenna Modifications:

Modifying the antenna of a Baofeng radio can significantly affect its performance and may lead to violations of FCC regulations. It is essential to use the provided or approved antennas for the specific radio model and avoid making any alterations that could lead to non-compliance.

Labeling and Identification:

Baofeng radios must display the appropriate FCC label and identification, including the FCC ID and compliance statements. This is something that should not be overlooked as it is essential to demonstrate compliance during inspections or inquiries.

Licensing Requirements for Amateur Radio Operators

As an amateur or Ham radio operator, you are able to access to a wide range of frequencies and communicate with fellow amateur operators globally. That being said, to operate a Baofeng radio for amateur radio purposes legally, you must obtain the appropriate FCC license.

In this section we break down the different types of amateur radio licenses:

Technician Class License

The Technician Class license is the entry-level license for amateur radio operators. With this license, you can operate on specific VHF and UHF frequencies, including those typically used by Baofeng radios. Here are the key points regarding the Technician Class license:

Requirements: There is no requirement for previous experience or knowledge of electronics. The Technician Class license exam consists of 35 multiple-choice questions, and you need to score at least 26 correct answers (74.3%) to pass.

Frequency Privileges: Technicians have access to specific portions of the 10-meter, 6-meter, and 2-meter bands, as well as certain frequencies on the 70-centimeter band.

Limitations: Technicians are not allowed to use frequencies below 30 MHz, preventing them from accessing HF (high-frequency) bands. Additionally, there are some restrictions on power output and the use of certain modes.

General Class License

The General Class license grants expanded privileges, including access to HF frequencies, which allow for long-distance communication. Obtaining a General Class license is a significant step from the Technician Class license.

Requirements: To obtain a General Class license, you must already hold a Technician Class license. The General Class exam consists of 35 multiple-choice questions, where you need to score at least 26 correct answers (74.3%) to pass.

Frequency Privileges: General Class operators have access to all the Technician frequencies and gain privileges on most HF bands.

Limitations: While General Class operators enjoy greater access to HF bands, there are still some frequency segments and modes reserved for higher license classes.

Extra Class License

The Extra Class license is the highest level of amateur radio licensing in the United States. This license provides the most extensive frequency privileges and allows for unrestricted access to all amateur radio bands.

Requirements: To obtain an Extra Class license, you must already hold a General Class license. The Extra Class exam consists of 50 multiple-choice questions, where you need to score at least 37 correct answers (74%) to pass.

Frequency Privileges: Extra Class operators have access to all amateur radio frequencies, including exclusive sub-bands on HF.

Limitations: The Extra Class license does not have any significant limitations, as it provides full access to all available amateur radio bands and modes.

Obtaining an Amateur Radio License:

Study: To prepare for the license exam, study the relevant material using online resources, books, or local ham radio clubs that offer classes.

Find an Exam Session: Locate an amateur radio license exam session near you. Exam sessions are often held by local ham radio clubs, public events, or examiners designated by the FCC.

Take the Exam: Attend the exam session and take the written exam for the desired license class. Exams are usually conducted in person, but remote options may be available in certain situations.

Submit Application: If you pass the exam, fill out the required application form and submit it to the FCC along with the appropriate fee.

Receive Your License: Once the FCC processes your application, you will receive your amateur radio license, complete with your assigned call sign. This call sign will identify you as a licensed operator on the airwaves.

Renewing and Upgrading Licenses

Amateur radio licenses are valid for 10 years from the date of issue and require renewal to remain active. The FCC provides ample time for licensees to renew before expiration. Additionally, as operators gain experience and knowledge, they may choose to upgrade their licenses by taking a superior level exam. To upgrade, one must file a renewal application with the FCC before the expiration date. The renewal process can be completed online through the FCC's Universal Licensing System (ULS) or by submitting a paper application to the FCC.

Responsible and Ethical Radio Use for Outdoor Enthusiasts

As outdoor enthusiasts and amateur radio operators, we not only have a legal obligation to adhere to the regulations governing radio communications but also a moral and ethical responsibility to use our radio devices in a respectful and considerate manner. Being amateur ratio operators is a privilege that allows us to enjoy and enhance our outdoor experiences, but it comes with an inherent duty to prioritize safety, respect others, and protect the environment. Let us delve into the moral and ethical aspects of using radio devices in outdoor activities:

1. **Safety First:** The primary moral duty of any radio operator is to prioritize safety. Whether we are hiking, camping, or engaging in other outdoor adventures, our Baofeng radios can be vital tools for communication during emergencies. Ensuring the safety of ourselves and others is of paramount importance. By being prepared, knowledgeable, and responsible in our radio use, we can contribute to safer outdoor experiences for everyone.

2. **Being Mindful of Others:** Ethical radio use involves being considerate of other users, both within the amateur radio community and beyond. Avoid monopolizing channels and always be mindful of the purpose of each frequency. During emergency situations, give priority to essential communications and avoid unnecessary chatter that may interfere with critical operations.

3. **Honesty and Integrity:** Upholding honesty and integrity in our radio communications is a fundamental ethical principle. Avoid misrepresenting your identity or purpose when communicating on the airwaves. Be truthful in all interactions and refrain from using your radio to deceive or spread false information.

4. **Respecting Privacy:** Respecting the privacy of others is a key ethical consideration. Radio transmissions are not private and can be intercepted by anyone with the appropriate equipment. As responsible radio operators, we must avoid sharing personal or sensitive information over the airwaves. Protect the privacy of yourself and others by exercising discretion in your communications.

5. **Preserving the Environment:** As outdoor enthusiasts, we share a deep love for nature and the environment. Ethical radio use involves minimizing our impact on the environment and wildlife. Refrain from using radios in ecologically sensitive areas or disturbing wildlife with excessive noise. Leave no trace and strive to protect the natural beauty of the outdoors for future generations.

6. **Avoiding Harm:** Ethical radio use requires us to avoid causing harm to others through our communications. Refrain from engaging in offensive, harassing, or malicious speech over the airwaves. Treat all fellow radio operators with respect and kindness, promoting a positive and inclusive radio community.

7. **Contributing to Public Safety:** Amateur radio operators have a unique role in supporting public safety and emergency communication. Embrace this responsibility ethically by staying informed about emergency protocols, participating in drills and exercises, and offering assistance when needed. Your Baofeng radio can be a lifeline for others in times of distress, and ethical operators are ready to step up when called upon.

8. **Continuous Improvement:** Ethical radio operators recognize that learning is a lifelong journey. Strive to continuously improve your skills, knowledge, and understanding of radio communications. Participate in training opportunities, engage with experienced operators, and be open to feedback to enhance your effectiveness and ethical practice.

9. **Encouraging Positive Behavior:** Be an ambassador for ethical radio use in the outdoor adventure community. Encourage others to obtain proper licenses, adhere to regulations, and follow responsible radio practices. By setting a positive example, we can foster a culture of ethical radio use and create a stronger, more united community.

10. **Promoting Diversity and Inclusion:** Embrace diversity and inclusivity in the amateur radio community. Welcome and support individuals from all backgrounds, regardless of gender, race, ethnicity, or any other characteristic. Cultivate an environment where everyone feels respected and valued.

By embodying these moral and ethical principles in our radio use, we not only comply with regulations but also contribute to a vibrant and responsible amateur radio community. Let us remember that ethical radio operators are not just operators of technology but also stewards of an invaluable resource that can connect and protect us during our outdoor adventures. Upholding these principles ensures that we can enjoy the full potential of our Baofeng radios while being responsible custodians of this essential form of communication.

Conclusion

I hope this handbook has proven to be a good source of information for you to explore the world of radio communication, the laws governing it, and the ethical duties of operators. Independently of what situation you find yourself in, whether you are hiking, camping or doing any other outdoor activity, having a reliable communication system is a lifeline. Communication systems like Baofeng radios become even more important during emergencies. Disasters and emergencies happen out of a sudden, and Baofeng radios are a great tool during such times.

Encouraging the Implementation of Knowledge for Safe and Effective Communication

Now that you have acquired the basis and the technical information, it's your turn to put this knowledge into action. But you need to be ready!

1. Practice regularly to become proficient. Familiarize with the controls, experiment with various antenna types, and explore the radio's features.
2. Obtain the Necessary Licenses from the relevant regulatory authorities.
3. Embrace responsibility. As an amateur radio operator you will have to adhere to the principles of ethical radio use. Promote good communication practices and offer assistance in emergencies.
4. Whether you are an outdoor enthusiast, prepper, or hobbyist, explore ways to integrate Baofeng radios into your activities.
5. Share what you have learned with fellow operators, beginners, and anyone interested in communication preparedness.
6. Technology and regulations in the radio world are constantly evolving. Keep an eye out for the latest developments, upgrades, and changes in frequency allocations. When doing so, make sure you rely on reputable sources.
7. Cultivate a resilient mindset that can adjust to changing circumstances.
8. While Baofeng radios are powerful tools, they should always be used responsibly and in accordance with local regulations.

Final Thoughts

Thank you for choosing The "Baofeng Radio Survival Handbook" as your guide to radio communication. I hope that this reading has provided you with the knowledge and insights you were expecting to find in this book.

Keep learning about your radio and continue testing their features and capabilities. But remember, acquiring the appropriate licenses and complying with FCC regulations is not only a legal requirement but also a demonstration of respect to other radio users.

With gratitude,

Alex Ranger

Appendix Glossary and Additional Resources

Glossary

AM (Amplitude Modulation): A method of encoding information onto a radio wave by varying its amplitude.

Amateur Radio: A personal radio service for non-commercial communication, experimentation, and public service.

Antenna: A device that transmits or receives radio waves. It converts electrical signals into radio waves for transmission or vice versa.

APRS (Automatic Packet Reporting System): A digital communication system that uses amateur radio frequencies for transmitting real-time data such as GPS coordinates and weather information.

Band: A range of frequencies allocated for specific radio communication purposes.

Bandwidth: The range of frequencies that a communication channel can support.

Baofeng Radio: A brand of handheld transceivers known for their affordability and versatility, popular among amateur radio operators.

Battery Eliminator: An accessory that allows a radio to be powered directly from an external power source, bypassing the need for batteries.

BNC Connector/Adapter: A type of coaxial connector commonly used in radio equipment.

CTCSS (Continuous Tone-Coded Squelch System): A sub-audible tone used to access a specific channel, reducing interference from other users on the same frequency.

Call Sign: A unique identifier assigned to licensed radio operators.

Channel: A specific frequency within a radio band assigned for communication.

Chirp: A popular software used for programming Baofeng radios.

Digital Mode: A mode of radio communication that uses digital signals rather than analog signals.

Dual-Band: A radio that can operate on two different frequency bands.

Duplex: A mode of operation where a radio can both transmit and receive on different frequencies simultaneously.

Electromagnetic Spectrum: The entire range of frequencies of electromagnetic radiation, including radio waves, microwaves, infrared, visible light, ultraviolet, X-rays, and gamma rays.

Emergency Frequency: A designated frequency used for emergency communication and distress calls.

FCC (Federal Communications Commission): The United States regulatory agency responsible for managing and licensing radio communication.

FRS (Family Radio Service): A license-free personal radio service with short-range communication capabilities.

Frequency: The number of cycles per second of a radio wave, measured in Hertz (Hz).

GMRS (General Mobile Radio Service): A personal radio service that requires a license and provides higher power and longer-range communication than FRS.

Ground Plane: A conducting surface used to improve the performance of an antenna.

Handheld Radio: A portable radio that can be operated while held in hand.

Ham Radio: Informal term for Amateur Radio.

High Frequency (HF): A range of radio frequencies between 3 MHz and 30 MHz.

Interference: Unwanted signals or noise that disrupt radio communication.

kHz (Kilohertz): One thousand Hertz, a unit of frequency commonly used in radio communication.

License: An authorization granted by a regulatory authority, such as the FCC, to operate a radio transmitter within specific frequency bands and power limits.

Low Frequency (LF): A range of radio frequencies between 30 kHz and 300 kHz.

Longwave: The lowest frequency range in the radio spectrum, between 30 kHz and 300 kHz.

Low Power Operation (QRP): Operating a radio with lower than standard power output for efficiency and portability.

LMR (Land Mobile Radio): A two-way radio system used for communication between mobile and stationary users, often used by public safety agencies and businesses.

Local Area Network (LAN): A network of interconnected devices within a limited geographical area, such as a home or office, that allows communication and resource sharing.

Long-Distance Communication: Radio communication over extended distances, typically achieved using repeaters, high-power transmitters, or atmospheric propagation phenomena.

Long-Term Evolution (LTE): A standard for wireless broadband communication, used for high-speed data transmission in modern cellular networks.

Low-Profile Antenna: An antenna designed to have a minimal physical footprint, suitable for inconspicuous installations.

LEO (Low Earth Orbit) Satellite: A satellite that orbits the Earth at relatively low altitudes, providing short communication delays and rapid signal transmission.

LNA (Low-Noise Amplifier): An electronic amplifier that amplifies weak signals without introducing significant additional noise.

LOS (Line of Sight): The direct visual path between two radio communication points, without obstructions such as buildings or mountains.

MHz (Megahertz): One million Hertz, a unit of frequency used in radio communication.

Modulation: The process of encoding information onto a radio wave by varying its amplitude, frequency, or phase.

Morse Code: A method of encoding text characters as sequences of dots and dashes, used in early radio communication.

Multipath Interference: Signal distortion caused by radio waves arriving at the receiver through multiple paths, resulting in phase cancellation.

NOAA (National Oceanic and Atmospheric Administration): A U.S. government agency that provides weather forecasts and warnings.

NWR (NOAA Weather Radio): A nationwide network of radio stations broadcasting continuous weather information from NOAA.

Narrowband: A mode of operation with a narrow bandwidth, enabling more channels to fit within the radio spectrum.

Oscillator: An electronic circuit that generates a stable frequency signal used as a reference for radio transmission.

Output Power: The power level at which a transmitter sends radio waves.

Packet Radio: A digital communication method where data is sent in short bursts, often used in amateur radio.

PTT (Push-To-Talk): A button or switch that activates the microphone for transmitting.

PL Tone (Private Line Tone): See CTCSS.

Programming Cable: A cable used to connect a radio to a computer for programming frequencies and settings.

QRZ: A popular website and online community for amateur radio operators.

QSL Card: A confirmation card exchanged by amateur radio operators to verify successful communication.

Radio Waves: Electromagnetic waves used for wireless communication, with frequencies between a few Hz and several GHz.

Repeater: A radio device that receives a weak signal and retransmits it at higher power to extend communication range.

RF (Radio Frequency): The range of frequencies used for radio communication.

Scan: The process of automatically searching and listening to multiple channels to find active signals.

Simplex: A mode of communication where radios transmit and receive on the same frequency.

SSB (Single Sideband): A mode of AM modulation that reduces bandwidth and improves long-distance communication.

Squelch: A circuit that mutes the receiver when no signal is present, reducing background noise.

Tone Squelch: See CTCSS.

Transceiver: A radio device that can both transmit and receive signals.

UHF (Ultra High Frequency): A range of radio frequencies between 300 MHz and 3 GHz.

VHF (Very High Frequency): A range of radio frequencies between 30 MHz and 300 MHz.

Watts: A unit of power used to measure radio transmitter output.

Weather Alert: A feature that automatically alerts users to severe weather conditions on NOAA weather radios.

Wideband: A mode of operation with a wider bandwidth, suitable for transmitting large amounts of data.

WSPR (Weak Signal Propagation Reporter): A digital mode used for weak signal communication and propagation studies.

Wavelength: The distance between two corresponding points on a radio wave, measured in meters or feet.

WiFi: A technology that uses radio waves to provide wireless internet access.

Wireless: Any form of communication that does not require physical wired connections.

XTAL (Crystal): A component used in some radio transmitters to generate a stable frequency.

Yagi Antenna: A directional antenna with multiple elements, used for high-gain and long-range communication.

Zello: A smartphone app that simulates push-to-talk communication using internet connectivity.

Z-Wave: A wireless communication protocol used for home automation and smart devices.

Additional resources for learning

Amateur Radio Relay League (ARRL): The ARRL is a prominent organization for amateur radio operators, offering a wealth of information, resources, and training materials. Their website provides articles, tutorials, and study guides for obtaining amateur radio licenses.

Baofeng Tech: The official Baofeng Tech website provides product manuals, software downloads, and firmware updates for Baofeng radios. It also offers customer support and troubleshooting guides.

HamStudy.org: This website offers free practice exams and study tools for aspiring amateur radio operators. It can help individuals prepare for their licensing exams effectively.

Hamradioacademy.com: This website provides links to all sorts of resources including pratical tips, study guides and reference charts.

YouTube Channels such has Ham Radio Crash Course (HRCC) which covers a wide range of ham radio topics, including Baofeng radio reviews, programming guides, and antenna setups.

You can find a lot of additional information by joining local amateur radio clubs, online forums and communities, Facebook groups, or subscribing to dedicated Podcasts. There are also plenty of publications and magazines for you to stay up to date with the latest developments.

Q-codes

Q-codes or Q-signals are an established system of radio shorthand which date back to the early days of wireless communication, with origins in early telegraphy codes. The codes where originally adopted by commercial radiotelegraph communication and developed to facilitate swift information exchange between operators of diverse linguistic backgrounds, these abbreviations have found extensive adoption in modern ham radio practices. The following table presents the most prevalent Q-signals widely employed by ham radio operators. While Q-codes were initially devised for Morse code operations, their integration has become pervasive even in voice communications. For instance, you might frequently hear phrases like "QRZed?" to inquire about the caller's identity, or "I'm experiencing a little QRM" when an operator faces interference. Moreover, the exchange "Let's QSY to 146.55" signifies a coordinated shift from a repeater frequency to a nearby simplex communications frequency.

Abbr.	Questions
QRG	What is your exact frequency or the frequency of _____?
QRL	Are you currently busy or occupied with _____? Typically used to check if a frequency is already in use.
QRM	Your transmission is experiencing interference at a level of _____. Is my transmission being interfered with as well?
QRN	I am experiencing static interference at a level of _____. Are you also troubled by static?
QRO	Should I increase power?
QRP	Should I decrease power?
QRQ	Would you like me to send faster at a speed of _____ words per minute?
QRS	Would you prefer me to send more slowly at a speed of _____ words per minute?
QRT	Shall I stop transmitting?
QRU	I have no messages for you. Do you have any messages for me?
QRV	Are you ready to communicate?
QRX	When will you call me again? (at _____ hours on _____ kHz). Minutes are usually implied rather than hours.
QRZ	Someone is calling you. Who is calling me?
QSB	Your signals are fading. Are my signals also fading?
QSK	Can you hear me between your signals, and can I break in on your transmission?
QSL	I acknowledge receipt. Can you also acknowledge receipt of my message or transmission?

QSO	I can directly communicate with _____ (or relay through _____). Can you communicate directly with _____ or by relay?
QSP	I will relay to _____. Will you relay to _____?
QST	This is a general call preceding a message addressed to all amateurs and ARRL members.
QSX	I am listening to _____ on _____kHz. Are you listening to _____ on _____kHz?
QSY	Shall we change to transmission on another frequency (or on _____kHz)?
QTC	I have _____ messages for you (or for _____). How many messages do you have to send? These Q signals are commonly used on the air. (Q abbreviations are in the form of questions only when they are sent followed by a question mark.)
QTH	My location is _____. What is your location?
QTR	The time is _____. What is the correct time?

Made in the USA
Las Vegas, NV
18 October 2023

79224385R00050